AGRICULTURAL USE OF GROUNDWATER

FONDAZIONE ENI ENRICO MATTEI (FEEM) SERIES ON ECONOMICS, ENERGY AND ENVIRONMENT

This series serves as an outlet for the main results of FEEM's research programmes in the areas of economics, energy and environment.

The titles published in this series are listed at the end of this volume.

Agricultural Use of Groundwater

Towards Integration Between Agricultural Policy and Water Resources Management

Edited by

CESARE DOSI

University of Padova and Fondazione Eni E. Mattei

KLUWER ACADEMIC PUBLISHERS
DORDRECHT / BOSTON / LONDON

Library of Congress Cataloging-in-Publication data is available.

ISBN 0-7923-6805-3

Published by Kluwer Academic Publishers
PO Box 17, 3300 AA Dordrecht, The Netherlands

Sold and distributed in North, Central and South America
by Kluwer Academic Publishers,
101 Philip Drive, Norwell, MA 02018, USA

In all other countries, sold and distributed
by Kluwer Academic Publishers,
PO Box 322, 3300 AH Dordrecht, The Netherlands

Printed on acid-free paper

Printed and bound in Great Britain by MPG Books Limited, Bodmin, Cornwall.

Table of Contents

Preface

Groundwater is endangered and polluted in several ways. Conservation and better management of this invisible resource should be a key ingredient of water policies. This is especially true in those areas, such as southern European regions, which are most affected by scarcity problems and competition between fresh water uses and users.

Agriculture is an important user of groundwater not only in terms of abstractions, but also in an incidental sense, by generating and/or releasing pollutants such as salts, nitrates, and plant-protection products, altering groundwater quality. Agricultural policies, traditionally directed towards other objectives, are beginning to pay more attention to environmental considerations. However, more effective initiatives are required to reduce the pressure upon groundwater resources and to achieve a better integration of agricultural and environmental policies.

This volume has developed out of three workshops held as part of the concerted action SAGA, 'Sustainable Agricultural Use of Aquifers in Southern Europe: Integration between Agricultural and Water Management Policies' (*FAIR5-CT97–3673*) carried out with financial support from the Commission of the European Communities, Agriculture and Fisheries (FAIR) RTD programme.

The workshops brought together SAGA partners, as well as other European scholars working in different but complementary fields. The aim was to get a picture of the interlinkages between agriculture, agricultural policies and groundwater management, to review policy approaches and instruments for improving management, and to identify further research directions.

I wish to thank all contributors and colleagues working on the behalf of SAGA partners who have actively contributed to the preparation of this volume. I am also grateful to three external referees for many helpful comments, and to Roberta Ranzini, Dino Pinelli and Francesca Carobba for their assistance during the process that took us to the final manuscript.

Finally, I wish to thank the Commission for funding the workshops and the preparation of this volume, whose content, however, does not necessarily reflect its views and in no way anticipates the Commission's future policy in this area.

Cesare Dosi

Contributors

Marc Barbier is a researcher in socio-economics and management sciences at the National Institute for Agronomic Research (INRA), Grenoble, France.

Nir Becker is a senior lecturer of environmental and resources economics in the Tel-Hai College (Israel), and a researcher at the Natural Resources and Environmental Research Center (NRERC), University of Haifa, Israel.

Hans Th.A. Bressers is a professor in policy studies and environmental policy and the scientific director of the Center for Clean Technology and Environmental Policy (CSTM), University of Twente, the Netherlands

Floor Brouwer is head of the Research Unit on Environment and Technology in the Agricultural Economics Research Institute (LEI), Wageningen University, the Netherlands.

Javier Calatrava is a doctoral student and research assistant in the Department of Agricultural Economics, Universidad Politecnica de Madrid, Spain.

Felisa Ceña is a professor in agricultural economics and policy in the Department of Agricultural Economics, Sociology and Policy, University of Córdoba, Spain.

Eduardo Chia is a senior researcher in agricultural economics at the National Institute for Agronomic Research (INRA), Dijon, France.

Cesare Dosi is an associate professor in environmental economics and public finance in the Department of Economics, University of Padova, and a senior research fellow at the Fondazione ENI E. Mattei (FEEM), Milano, Italy.

Alberto Garrido is an associate professor in environmental and resource economics and microeconomics in the Department of Agricultural Economics, Universidad Politecnica de Madrid, Spain.

Andrea Giacomelli is a researcher in hydrologic modelling and GIS at the Center for Advanced Studies, Research and Development in Sardinia (CRS4), Cagliari, Italy.

Carlo Giupponi is a lecturer in the Department of Agronomy, University of Padova, and a senior researcher in land use and natural resources management at the Fondazione ENI E. Mattei (FEEM), Milano, Italy.

Ingo Heinz is a senior economist at the Institute for Environmental Research (INFU), University of Dortmund, Germany.

Nuria Hernández-Mora is researcher in land use and water resources management at the Fundación Marcelino Botín, Madrid, Spain.

Giannis Karagiannis is a lecturer in agricultural economics and microeconomics in the Department of Economics, University of Crete, Greece

Stefan M.M. Kuks is a senior research associate in the Center for Clean Technology and Environmental Policy (CSTM), University of Twente, and executive member of the Regge and Dinkel Waterboard, the Netherlands.

Ramón Llamas is a professor of hydrogeology in the Universidad Complutense de Madrid, Spain, and a member of the board of directors of the International Water Resources Association.

Luis Martínez Cortina is a researcher in groundwater management and modelling at Fundación Marcelino Botín, Madrid, Spain.

Dionisio Ortiz is a doctoral student and research assistant in the Department of Agricultural Economics, Sociology and Policy, University of Córdoba, Spain.

Claudio Paniconi is a senior researcher in hydrology and environmental modelling and head of the Environment Division of the Center for Advanced Studies, Research and Development in Sardinia (CRS4), Cagliari, Italy.

Geerten J.I. Schrama is a senior research associate in the Center for Clean Technology and Environmental Policy (CSTM), University of Twente, the Netherlands.

Stefania Tonin is a researcher in environmental economics and water resources management at the Fondazione ENI E. Mattei (FEEM), Milano, Italy.

Anastasios Xepapadeas is a professor of economics in the Department of Economics, University of Crete, and a senior research fellow at the Center for Economic Research and Environmental Strategy (CERES), Athens, Greece.

Naomi Zeitouni is a lecturer of environmental and resources economics in the Department of Economics and a researcher at the Natural Resources and Environmental Research Center (NRERC), University of Haifa. Israel.

PART I

Introduction

CHAPTER 1

Introduction – Improving Agricultural Use of Groundwater: Policy Issues and Further Research Directions

Cesare Dosi

1.1. Background

Despite its multiple functions, groundwater has suffered from undervaluation and neglect. However, in recent years, which have witnessed a gradual shift from the traditional 'supply-side' approach (increasing availability/accessibility) towards a more balanced strategy which emphasizes 'demand-side' management and conservation of fresh water resources, more attention has been paid to better management of groundwater.

In 1996, the Commission released the proposal *An Action Programme for Integrated Groundwater Protection and Management*. In this proposal, the emphasis was put on the proper management of groundwater as a key component of Member States' water policies and, within the overall objective of groundwater conservation, the giving of priority to relieving the pressure exerted by diffuse sources (European Commission, 1996). The necessity for this is highlighted in a Commission communication on the state of Europe's environment: whilst there have been substantial improvements in surface water quality due to reductions in point source discharges, pollutant releases from agricultural non-point sources have shown little change, and the maximum admissible concentrations of nitrate and certain pesticides are frequently exceeded (European Commission, 1999).

According to the 1996 Commission's proposal, a sustainable quality management should protect and preserve all groundwaters, and the actions to achieve this objective should be based on prevention, action at source, and the polluter pays principle (European Commission, 1996). It is to be hoped that this somewhat ambitious programme and the proposed general action lines will be translated into effective policy initiatives. While the lack of initiatives aimed at improving resources management, from both a quantitative and qualitative point of view, would negatively affect all countries in the long run, the European areas which will suffer the most from the deficit of adequate policy measures

C. Dosi (ed.), Agricultural Use of Groundwater, 3–11.

will undoubtedly be those where scarcity problems and competition between alternative water uses and users are relatively more acute and widespread.

In quantitative and global terms, southern Member States benefit from a relative abundance of renewable fresh water resources. However, although national data (based on long-term average annual precipitation and evapotranspiration rates) tend to provide a reassuring picture, they hide significant geographical imbalances. Moreover, southern European countries, and in particular semi-arid and arid areas, are characterized by a significant seasonal and interannual variability in precipitation rates. Finally, national statistics on resources endowments usually tell little (or nothing) about the qualitative status of the theoretically exploitable water resources.

As far as abstractions are concerned, with the exception of France, agriculture is the most important water user in southern Europe: although the breakdown of consumption between the various sectors varies considerably from one region to another, water abstracted for irrigation in Greece, Spain, Italy and Portugal accounts for more than 60% of total consumption, while in the rest of Europe, on average, agriculture contributes to less than 10% of total estimated withdrawals. As far as groundwater is concerned, with the exception of France and Portugal (about 17% and 23%, respectively), agriculture is the most important user of groundwater: in Greece, Italy and Spain, agricultural use accounts for about 58%, 57% and 80% of total groundwater abstractions, respectively.

However, for agricultural use of inland waters in general and groundwater in particular, besides *direct uses* (withdrawals), which may imply various short or long-term third party effects, effects which stem from quantity depletion, quality degradation (e.g. salt water intrusion in coastal regions), or off-site externalities such as subsidence and loss of natural habitats, special attention should also be paid to *incidental uses*. The latter includes farming practices which entail, as a by-product, the generation of pollutants which have the potential to have a significant impact on the status of groundwater. Although the information is patchy, the concern about the increasing concentrations of nitrates and pesticides in several regions, which often exceed the EU maximum admissible concentrations, is legitimate.

Groundwater is an important source of water supply. With the exception of Spain (22%), the sector which is most dependent on groundwater is the domestic one: in Italy, Greece and France, the relative contribution of groundwater to total domestic supplies is in the area of 91%, 64% and 57%, respectively. In addition, besides being a major source of fresh water, groundwater provides important buffers against the seasonal mismatch between water supply and demand: this buffering action, particularly valuable in semi-arid and arid areas, is likely to become even more important in the future. Moreover, the importance of groundwater is also related to its contribution to waste assimilation, to the flow of streams and rivers, and to its support of wetlands and aquatic ecosystems.

All these features should be taken into proper account when assessing whether current exploitation patterns are consistent with a 'sustainable' use of aquifers. In principle, this assessment should be aimed at comparing the benefits and costs of groundwater exploitation from a social perspective, and in the presence of negative indications, at identifying appropriate changes with respect to current exploitation patterns. As with other economic sectors, it is reasonable to believe that the net social value of farmers' use of groundwater will often be negative. This is especially true in those contexts where, because of the failure to regulate private (direct and/or incidental) uses, aquifers are de facto 'open-access' resources, i.e. resources whose exploitation (either as a source of water or as a sink for pollutants) does not entail a private cost, or entails a cost which does not reflect the true short and long-term social costs of aquifer exploitation.

Although reconciliation between private uses of natural resources and social objectives generally requires the implementation of ad hoc environmental policy measures, in many instances, better resources management could be also achieved through reforming existing sector policies by removing distorting incentives or integrating environmental objectives into policies traditionally designed to achieve other social goals.

This is especially true for the European agricultural sector and for the Common Agricultural Policy (CAP). Until very recently, rather than promoting a more socially efficient use of natural resources by farmers, agricultural policies have often added further distortions, and by so doing, worsened resource misallocation. The need to integrate environmental objectives into CAP has been acknowledged by European authorities, and has influenced recent reforms to some extent. However, there is a legitimate concern that the achieving of a better integration of agricultural and environmental policies and a more fruitful division of labour will require more effective initiatives.

1.2. Main results: policy recommendations and further research directions

On the basis of a literature and policy review, the exchange of information between partners, and analyses and recommendations proposed in the following chapters, the main results and conclusions of this Concerted Action can be summarized as follows:

- Southern European countries, in spite of their relative global abundance of water resources, are afflicted by significant internal imbalances. A proper understanding of the nature, causes, and intensity of local scarcity problems is obviously a prerequisite for any policy aimed at removing or attenuating present imbalances. However, one of the main difficulties encountered when trying to obtain a reliable picture about scarcity problems is that official water statistics tend to use aggregated indicators and do not provide adequate and comprehensive information about the qualitative status of available resources. Despite these caveats, it is known that many southern regions are afflicted by occasional or structural water stresses. In these areas, physical scarcity (often exacerbated by poor management and inadequate regulation of private uses) generates tension and conflicts, and often represents an inhibiting factor for economic development. Although the information available is in many cases anecdotal, there is a widespread perception that in many areas, competition between water users and uses is increasing.

- Proper management of groundwater is a key component of any sustainable water management policy. Sustainable groundwater management requires special attention to be devoted to agricultural activities. This not only because agriculture is a large user in terms of volume, but also because it is often a pervasive cause of quality problems. These problems, which may be traced back to over-pumping and/or to the use of potentially polluting inputs, include salt water intrusion in coastal aquifers, salinization due to the solubilization and percolation of salts previously deposited along the soil profile, and contamination by pesticides and nitrate leaching. Although these phenomena also affect other European countries, some of them are particularly significant in some southern regions, or in any case tend to produce more serious consequences due to the relatively more acute scarcity problems that afflict these areas.
- A key component in any definition of sustainable water use is a better understanding of the concept of aquifer 'over-exploitation'. A number of terms related to overexploitation appear in the literature and in policy documents. These include safe yield, sustained yield, overdraft, exploitation of fossil groundwater, and optimal yield. All these terms have in common the idea of avoiding undesirable effects as a result of intensive groundwater development. However, this undesirability depends mainly on the social perception of the issue, and this is more related to the cultural, regulatory, and economic context than to strict hydrogeological data. Aquifer overexploitation is a complex concept that needs to be understood in terms of a comparison of the social, economic, and environmental benefits and costs that derive from a certain level of groundwater (direct and/or incidental) use.
- Estimating these benefits and costs, and, more generally speaking, assessing the social value of groundwater, is important for both allocation decisions among different users as well as long-term decisions with respect to investments in restoration, conservation, and the development of alternative supplies or demand management decisions. The value of aquifers and the long-term analysis of groundwater use can be strongly correlated to the problem of sustainability. It could be said that a society is not moving along a sustainable path if the decline of groundwater quantity and quality is not compensated for by more effective and efficient water services, and/or if the costs of water services required to relax physical resources limits continuously increase over time, or if there are political or social constraints which impede adoption of institutional reforms required to make water resources allocation less socially inefficient.

- A proper evaluation of groundwater resources requires a proper identification of the services provided by an aquifer. Aquifers do not merely provide 'extractive values', but also 'in situ values'. These include stock value, buffer value, avoidance of seawater intrusion, option value, subsidence avoidance, and ecological values. The international literature provides a large variety of techniques (and empirical studies) aimed at estimating these components of the total value of an aquifer. Estimating this value is not easy, nor should it be interpreted as a clear-cut tool for guiding policy decisions in respect of resource allocation between users and uses, and conservation measures. However, a number of useful insights could be gained through economic assessment. In this respect, it is worthwhile noting that a review of the literature shows that the vast majority of available empirical studies have been conducted outside the European Union. Consequently, more research efforts on evaluation of groundwater uses (and abuses) should be promoted and carried out, this also in the view of the future implementation of the provisions of the Water Framework Directive (COM(97)49), proposed by the Commission, subsequently amended (COM(97)614, COM(98)76, COM(99)271), and currently being negotiated by the European Parliament and the Council of Ministers. As far as water pricing is concerned, the draft directive endorses the principle of full-cost recovery. Although the document does not provide an unambiguous and clear-cut operational translation of the full-cost principle, it has not ruled out the incorporation of scarcity values and environmental externalities in full cost recovery (OECD, 1999). Consequently more effort should be made to assess these values and externalities within the European Union.
- Whenever groundwater exploitation is believed not to be sustainable (that is, when the social benefits of groundwater exploitation are lower than the associated social costs), appropriate measures for altering private exploitation patterns are required. This is often the case in agriculture, where farmers often over-exploit groundwater, either through over-pumping or by incidentally using aquifers as a sink for pollutants. Measures aimed at affecting farmers' behaviours may take different forms, forms that may include mandatory regulation, economic incentives and so-called voluntary approaches. However, the effectiveness of these measures relies upon a proper understanding of the main features of the interlinkages between farming practices and groundwater quality. For instance, most of the agriculture-related groundwater pollution problems may be labelled as non point-source (NPS), which typically involve many agents, geographically dispersed, generating (and/or causing the intrusion into aquifers of) pollutants which, in general, cannot be easily be neutralized ex-post through end-of-pipe devices.
- Two main implications for policy design stem from the underlying features of NPS pollution. First, a preventive approach (avoidance/reduction of pollutant discharges) is the preferable, and sometimes the only viable option for controlling groundwater pollution from diffuse agricultural sources.

Second, the effectiveness (and efficiency) of a preventive approach can be undermined by the difficulty/impossibility of monitoring on-site emissions, or inferring individual responsibilities from observable total off-site discharges. Different regulatory strategies and policy instruments for addressing and overcoming these monitoring problems have been proposed. A prerequisite for all these strategies is a proper understanding of pollutant generation and transport processes, and in this respect, an important role could be played by environmental modelling explicitly targeted towards the supply of the information required for controlling groundwater uses (or abuses).

- Generally speaking, policy provisions aimed at controlling (water) pollution from agricultural sources have relied, and still largely rely upon what is sometimes referred to as 'voluntarism', but which can probably be better described as a 'soft persuasion though-subsidisation' approach. Besides being in conflict with the polluter pays ethics, this approach has not brought about a significant or widespread reversal of pollution trends. This ineffectiveness can be at least partly attributed to the somewhat ambiguous distinction between farmers' environmental services and environmental damage, a distinction which is supposed to provide the legal basis for deciding whether or not farmers are eligible for compensation for environmentally friendly adjustments. In various EC policy documents, environmental services (*target levels*, according to the Commission's terminology) are defined as the outcome of any environmentally friendly adjustment of farming which goes beyond the basic standards of environmental care (*reference levels*). As with any other politically constructed property rights systems, the conventional borderline envisaged by the Commission is obviously questionable. What matters, however, is that to be credible and operative, the adopted legal borderline requires a rigorous and unambiguous definition of the reference level in order to assess farmers' compliance with legal regulations, and to have a benchmark for identifying farmers' environmental services to be compensated by society. However, in the EU in general, and particularly in those Member States which have not properly identified and credibly imposed basic standards of environmental care (e.g. failure to implement the Nitrate Directive), the in some ways intrinsically ambiguous distinction between farmers' negative and positive environmental externalities has reinforced the attitude among farmers that they should wait for compensation for any environmentally friendly adjustment of farming. Such a consolidated attitude could make the implementation of cross-compliance measures introduced by Agenda 2000 (which, in principle, could partly bridge the gap between subsidization of farming activities and farmers' environmental performances) politically difficult.
- There is a clear need for better and more effective integration and co-ordination between CAP, environmental policies and national water resources management strategies. CAP measures must complement and

should be complemented by environmental measures and regulations. The efficiency and effectiveness of CAP provisions and of other measures as the basis of a clearer European framework specifying the principles of a division of labour between economic incentives and regulation related to positive and negative externalities of agricultural production, need to be assessed. In the gap between minimum environmental standards (the reference levels) and the sort of environmental services for which payments are made, there is a range of societal expectations about the responsibilities that farmers should have in respect of the environment. In order to arrive at a concrete formulation of the environmental and groundwater conservation conditions that have to be fulfilled, a proper definition of *good agriculture practices* (GAPs) is essential. GAPs may then become a benchmark for deciding whether a farmer is eligible for income support (e.g. in the context of cross-compliance).

- Besides better local targeting of CAP general provisions and integration/coordination between agricultural policy and environmental regulation, water resources management could be improved through negotiated agreements between farmers and water authorities or water companies. In northern Member States such as Germany and the Netherlands, several instructive examples of the potential advantages of co-operative agreements compared with traditional regulatory approaches may be found. However, with the possible exception of France, such agreements are rare in southern Member States, or do not exist at all. The cultural, institutional and economic reasons for this, and the question of under which conditions such agreements could be established, are of special interest and should deserve more attention when research priorities are identified.
- Investigation into research on the environmental impacts of the CAP shows a northern bias in the research coverage, with the majority of studies and research projects focusing on northern countries, and much fewer for southern. Research also puts stronger emphasis on temperate rather than Mediterranean crops, with water pollution problems by nitrates and pesticides reasonably well covered, but not groundwater supply problems and over-abstraction. Finally, there is a strong emphasis on research into agri-environmental measures (namely those introduced through Regulation 2078/92) in comparison to other components of the CAP. Although important, these measures (which remain a minor component of the CAP: around 4% of the total budget), may draw attention away from the bigger picture.

1.3. A guide to the volume

Chapter 2 provides an overview of freshwater availability and groundwater use in southern Europe (France, Greece, Italy, Portugal and Spain). The authors briefly illustrate the geographical variability of freshwater supplies, and provide information about water abstractions by source and general groundwater management provisions in southern Member States.

Chapter 3 describes the interlinkages between agricultural production processes and groundwater quality, and the basic mechanisms and dynamics involved in the flow of water and transport of contaminants in soil and aquifer systems. The authors provide examples of different modelling approaches, and their advantages and limitations are discussed and illustrated.

Chapter 4 suggests a taxonomy of groundwater services and provides an overview of valuation methods approaches aimed at assessing the economic value of these services and the social costs of groundwater mismanagement.

Chapter 5 also stresses the complex nature of groundwater services and the numerous socio-economic and ecological benefits derivable from groundwater use and resources conservation, benefits which should be taken into proper account when sustainable exploitation patterns are considered. The authors emphasize that most of the difficulties encountered in the design of policies aimed at avoiding over exploitation stem from uncertainties and the lack of adequate information about resource endowments, both from a quantitative and qualitative point of view.

These background contributions, which are mostly aimed at identifying the rationale behind improving agricultural management of groundwater, are followed by chapters dealing with policy initiatives aimed at reducing the pressure exerted upon groundwater quality by agricultural activities, and analyses of the impacts exerted upon water use by agricultural policies.

Chapter 6 provides an overview of policy instruments aimed at controlling groundwater pollution from agriculturally diffuse sources. These instruments are classified according to the way in which pollution control operates, that is, through introducing compulsion to the farmers' choice domain (command and control measures) or through affecting the pros and cons of alternative courses of action legally open to farmers (economic instruments and voluntary approaches), as well as according to the way in which the monitoring problems that arise from the nature of NPS pollution are addressed. The review proposed by the authors also includes a description of instruments introduced through recent reforms of CAP, instruments which, however, are more deeply analysed and discussed in Chapters 10, 11 and 12.

Chapters 7, 8 and 9 provide examples of negotiated agreements (voluntary approaches) between water authorities (or water supply companies) and farmers operating within or near groundwater catchment areas. These agreements, which could usefully supplement other policy measures, are either quite rare or non-existent in southern Member States. It is possible that the experiences gained in countries such as Germany and the Netherlands could be used to create the prerequisites for a more widespread application of these voluntary approaches. Chapter 7 illustrates German experiences in this field. Chapter 8 provides an overview of the opportunities provided by negotiated agreements on the grounds of experiences undertaken in Great Britain, the Netherlands,

Germany and the USA. Chapter 9 illustrates a negotiated agreement undertaken in France between a private water company and farmers operating in a groundwater catchment area to reduce nitrate concentrations. The authors illustrate the role played by the various actors involved in the negotiation process, and, in particular, the role played by experts working on behalf of public research institutions.

Chapters 10, 11 and 12 focus more specifically on the impacts of agricultural policy and CAP provisions by emphasizing the need of an integrated analysis of agricultural, environmental and water resources management policies. Chapter 10 provides a theoretical model which illustrates the interaction between agricultural policies and water use under conditions of production uncertainty and uncertainty of resource endowments at the hand of a theoretical model. The impact of various policy instruments on water abstractions and pollutant discharges from farmland is analysed, and the co-ordination of agricultural and environmental policies is examined.

Chapter 11 illustrates the possible effects of various agricultural policy scenarios on farmers' demand for water with reference to Spain. The authors also determine the extent to which agricultural policies could mitigate the income effects of water pricing policies in line with the EU Water Framework Directive as it has currently been drafted.

Chapter 12 explores the interlinkages between agriculture and groundwater resource management at the hand of experiences in southern Spain, and describes the joint impact on farmers' decisions exerted by CAP measures and water institutions.

Finally, Chapter 13 provides an overview of the way the environmental effects of CAP are currently being viewed, and identifies some research gaps and suggestions for further research. The research gaps identified by the author (on the grounds of experience gained through previous overviews of European research initiatives) substantially coincide with those which have been identified through the concerted action SAGA. In particular, as far as the interlinkages between agriculture, agricultural policies and water resources management are concerned, there is a northern bias in the research coverage, with pollution problems by nitrates and pesticides reasonably well covered, but not groundwater quantity management and over-abstraction.

References

European Commission, 1996, Proposal for an Action Programme for Integrated Groundwater Protection and Management. COM(96)315 Final.

European Commission, 1997, Proposal for a Council Directive Establishing a Framework for Community Action in the Field of Water Policy. COM (1997) 49 Final.

European Commission, 1999, Europe's Environment. What Directions for the Future? COM(1999) 543 Final, Brussels.

OECD, 1999, Agricultural Pricing in OECD Countries. OECD – Environment Directorate, ENV/EPOC/GEEI (98) 11/Final. www.oecd.org/env/docs.

PART II

Groundwater Use and the Value of Conservation

CHAPTER 2

Freshwater Availability and Groundwater Use in Southern Europe

Cesare Dosi and Stefania Tonin

2.1. Introduction

Competition between freshwater uses and users is increasing all over the world, although its causes, nature and intensity vary across countries, depending on their latitude, stage of development, and social perception of welfare gains/losses stemming from alternative water uses (and abuses). While in relatively water-abundant and wealthy communities, competition is gradually taking on the form of rivalry between consumptive (abstractions and wastewater disposal) and non-consumptive uses (e.g. recreational uses), in water-scarce areas, competition still primarily results from the lack of sufficient water of adequate quality for basic human requirements.

In countries and regions surrounding the Mediterranean basin, both forms of competition can be observed. However, while water scarcity in European countries tends to be localized, although rivalry between the demands for abstractions and resources conservation is gradually spreading, countries on the southern shores are experiencing a continuous global decrease in their ability to satisfy (traditional) basic needs.

The idea that to alleviate problems arising from global or local scarcity major changes in the approach to water resources management are required is gaining an increasing consensus, although it is still far from being translated into effective policy initiatives. One of the major changes required is a shift from the traditional preoccupation with increasing availability/accessibility, towards a more balanced approach which emphasizes demand-side management and conservation of freshwater resources.

When considering long-term sustainable management, special attention should be devoted to groundwater which, despite its multiple functions, has suffered from undervaluation and neglect. For instance, besides being a major source in many Mediterranean areas, aquifers provide important buffers against variations in water demand and supply: groundwater is stored underground in a natural way and does not require the construction of storage facilities, and

C. Dosi (ed.), Agricultural Use of Groundwater, 15–34.

it is cheaper to develop because it requires little if any prior treatment if it is not polluted (RIVM/RIZA, 1991). Moreover, the importance of groundwater stems from its contribution to waste assimilation, natural reticulation and sustenance of related stream flows and wetlands (Calow, 1998).

As far as abstractions are concerned, agriculture is the most important water user in southern Europe with the exception of France. Although the breakdown of consumption between the various sectors varies considerably from one region to another, according to available estimates (OECD, 1997; EEA, 1998; European Commission, 1999), water abstracted for irrigation in Greece, Spain, Italy and Portugal accounts for more than 60% of total consumption, whilst in the rest of Europe, on average, less than 10% of the resources are used for irrigation (European Environmental Agency, 1998b).

However, when considering agricultural use of inland waters in general, and groundwater in particular, besides *direct uses* (abstractions), which may imply quantity depletion, quality degradation (e.g. salt water intrusion in coastal regions), or off-site externalities such as subsidence and loss of natural habitats, special attention should be paid also to *incidental uses*, i.e. to farming practices which imply the generation of pollutants having the potential to have a significant impact on groundwater quality. For instance, although the information is patchy, there is a legitimate fear about increasing concentrations of nitrates and pesticides in several regions, concentrations which frequently exceed the EU maximum admissible concentrations.

The aim of this chapter is to provide an overview of freshwater availability and groundwater use in southern Europe. We begin with a description of the present situation and the prospects for the whole Mediterranean basin, and then we concentrate on EU Member States (France, Greece, Italy, Portugal and Spain), by focusing on the geographical variability of freshwater supplies, and on groundwater availability and management.

2.2. Freshwater availability and abstractions in the Mediterranean basin

2.2.1. *The present situation*

Generally speaking, water management problems are difficult to assess because of the shortage of adequate information. Besides the lack of reliable data about the resources available and abstractions, and this is especially true for agriculture where consumption is generally unmonitored, assessment of the nature and extent of scarcity problems suffers from the format in which statistics are available. Available statistics are usually provided on a national basis. However, it is well known that, *coeteris paribus*, the more uneven the geographical distribution of resources (and abstractions), the greater the risk of local shortfalls and internal competition for the allocation of national resources. In addition, estimates of freshwater potential supplies are based on annual and long-term average precipitation and evapo-transpiration rates. However, an area which seems relatively well supplied in terms of average parameters is not guaranteed protection against inter-annual fluctuations. Moreover, the probability of experiencing scarcity problems depends on the intra-annual distribution of water supplies: the more the average annual runoff is unevenly distributed throughout the year, the higher is the risk of shortfalls, especially when the (occasional or structural) mismatch between seasonal supplies and demands cannot be smoothed through artificial or natural buffers (reservoirs or accessible local aquifers). Finally, statistics on available resources do not usually provide adequate information about their qualitative status.

Despite the aforementioned caveats, indicators such as the annual average renewable water resources per capita or the ratio between (estimated) withdrawals to available renewable water resources can still constitute rough indexes of potential vulnerability to water shortages at a national scale.

Following Brouwer and Falkenmark (1989), renewable freshwater available in a country (TR) is defined as the total amount moving in rivers or aquifers. TR may be divided into the amount originating from domestic rainfall (annual internal renewable water flow, INT), or by water received from neighbouring countries in transboundary rivers and aquifers (annual inflows, INF). The ratio INF/TR is a rough indicator of a country's dependence on shared resources.

In Table 2.1, two indexes are presented: total renewable freshwater per capita (TRpc) and the ratio between total withdrawals and the annual internal renewable water flow (W/INT). To evaluate the present situation according to the first indicator (TRpc), it is worth remembering that, according to the Falkenmark's water stress index, while a country with more than 1700 m^3/year/person is expected to experience only intermittent and localized water shortages, the threshold of 1000 m^3 has been proposed as an approximate benchmark below which a country is likely to experience widespread and chronic shortfalls; at less than 500 m^3/person water availability becomes a primary constraint on socio-economic development.

In the countries surrounding the Mediterranean basin, the TRpc index exhibits a wide variability: it ranges from more than 10 000 m^3 to less than 50 m^3; for seven countries (Malta, Libya, Jordan, Israel, Tunisia, Algeria and Egypt) the TRpc is below (sometimes well below) 1000 m^3 (see Figure 2.1). If we look at the second indicator (W/INT), i.e. the ratio between withdrawals and annual renewable domestic resources, in four countries (Egypt, Libya, Malta and Israel), the index is significantly greater than or close to one (Tunisia): this means either that these countries mostly rely upon transboundary water flows (Egypt and Israel) or internal non-renewable and/or non-conventional sources (Libya, Malta and Israel).

Although abstractions from surface waters represent the largest share of total withdrawals, with the exception of some Middle East and African countries (Libya, Tunisia, Algeria and Israel), groundwater is an important source for domestic uses, agriculture and industry. The largest groundwater consumer is

Table 2.1. Renewable freshwater resources and withdrawals

	Annual internal renewable water flow (INT)		Annual inflows (INF)		Total renewable water resources (TR)		Annual withdrawals (W)				Sector withdrawals (%)		
	Total (km^3)	Per capita (m^3)	Total (km^3)	Per capita (m^3)	Total (km^3)	Per capita (m^3)	Year	km^3 (%)	W/INT (%)	W/TR	Domestic	Industry	Agriculture
Albania[1]	44.50	12,917	11,3	3,280	55.8	16,197	1970	0.20	0,4	0,36	6	18	76
Algeria[1]	13.87	460	0,4	13,26	14,27	473	1990	4.50	31	32	25	15	60
Cyprus**	0.90	1,213	0,0	0.0	0.90	1,212	1995	0.2	22	22	24	2	74
Croatia*	16	3,560	127	28,260	143	31,820	1990	–	–	–	–	–	–
Egypt[1]	2.8	43	55.5	845	58.3	888	1993	55.10	1968	95	6	8	86
France[1]	180	3,065	18	306	198.0	3,371	1990	37.73	21	19	16	69	15
Greece[1]	45.15	4,279	13.5	1,279	58.65	5,559	1980	5.04	11	9	8	29	63
Israel[2]	1.7	289	0.5	89	2.2	374	1989	1.85	109	84	16	5	79
Italy[1]	159.4	2,785	7.6	133	167.0	2,917	1990	56.20	35	34	14	27	59
Jordan[2]	1.3	218	0.4	67	1.7	285	1975	0.45	35	26	29	6	65
Lebanon[2]	5	1,565	0.6	188	5.6	1,753	1975	0.75	15	13	11	4	85
Libya[1]	0.6	100	0.0	0.0	0.6	100	1994	4.60	767	767	11	2	87
Malta**	0.0155	42	0.0	0.0	0.0155	42	1995	0.056	361	361	87	1	12
Morocco[1]	30	1,071	0.0	0.0	30.0	1,071	1992	10.85	36	36	5	3	92
Portugal[1]	38	3,878	31.6	3,225	69.6	7,103	1990	7.29	19	10	15	37	48
Slovenia*	19	9,901	14	7,295	33	17,196	1990	–	–	–	–	–	–
Syria[2]	25.8	1,682	27.9	1,819	53.7	3,502	1976	3.34	13	6	7	10	83
Spain[1]	110.3	2,775	1	25	111.3	2,800	1991	30.75	28	28	12	26	62
Tunisia[1]	3.52	371	0.6	63	4.12	434	1990	3.07	87	75	9	3	89
Turkey*	227	3,560	7	110	234	3,700	1995	35.10	15	15	–	–	–

[1] Data elaborated from World Resources Institute (1998–99).
[2] Data elaborated from World Resources Institute (1996–97).
* Data relating to Croatia, Slovenia and Turkey taken from EEA (1998a).
** Data relating to Malta and Cyprus taken from AQUASTAT, FAO (1995).

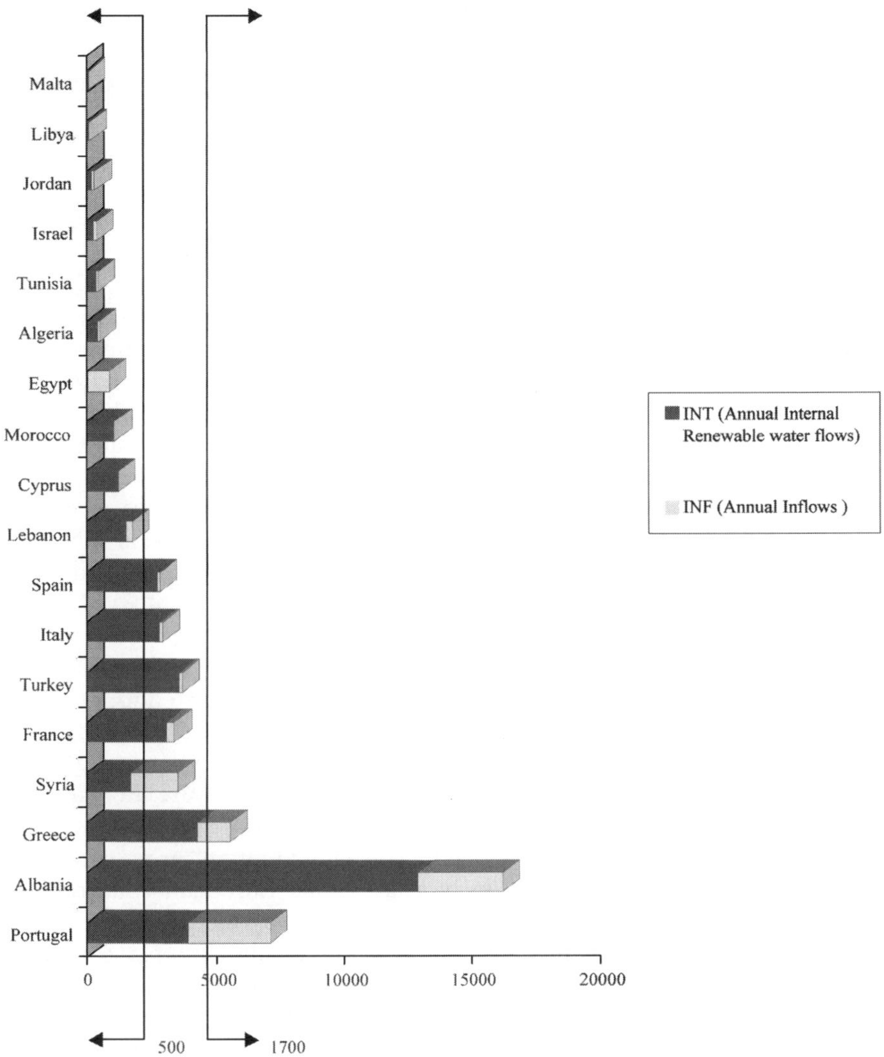

Figure 2.1. Total renewable freshwater resources (TRpc) (m^3/year/person).

Italy (more than 10 km^3/year), followed by Turkey, Spain and France (Margat, 1999). In many Mediterranean areas, abstractions exceed the average recharge rates, thus causing a continuous lowering of the water table levels; and there are countries (such as Libya, Malta and Tunisia) where there is a global imbalance between withdrawals and the total annual recharge of national aquifers.

In global terms, agriculture is the most important user of groundwater (with the exception of France, Malta and the former Yugoslavia): in Spain, for example, agriculture uses 80% of the total abstracted groundwater. However, if we consider the relative weight of different sources, the sector which is generally more dependent on groundwater supplies is the domestic one: in the large majority of Mediterranean countries, aquifers provide more than 50% of the total water directly consumed by households (91% in Italy, 64% in Greece, 57% in France) (Margat, 1999).

2.2.2. *General trends and scenarios*

Demographic pressure

When considering future water consumption scenarios, a distinction should be made between basic water requirements and water demands which include a much larger set of wants for water to provide additional goods and services (Lundqvist and Gleick, 1997). Demands for water, which, besides natural conditions, reflect countries' economic structures and institutional endowments, may be altered through appropriate policy initiatives and technological changes, without necessarily diminishing social welfare. For instance, since 1990 there has been a decline in total abstractions in many European countries. These can be partly attributed to more effective demand management which has reduced losses and used water more efficiently (EEA, 1998b).

It follows that it is hard to make predictions about future water demand patterns, since they will not only reflect demographic trends, but also developments in economic structures and the willingness and ability to adopt effective demand management strategies. Nonetheless, it is legitimate to believe that water demand will significantly increase in the Mediterranean basin as a whole, and this process will be mainly driven by demographic pressure. For instance, without taking migratory flows into consideration, according to current demographic projections, all southern Mediterranean countries are expected to experience a spectacular population increase, while European countries will experience either a slight increase or even a decrease by the year 2025.

The impacts of demographic trends on water availability per capita are illustrated in Table 2.2. Assuming that total annual average renewable water resources (i.e. domestic supplies and water received from neighbouring countries in transboundary rivers and aquifers) remain the same, all southern countries will experience a significant decrease in available resources per capita (TRpc), ranging between about 2526% (Israel) and 6053% (Libya). As a result, by 2025, the number of Mediterranean countries with a per capita water availability lower than 1000 m^3/year will have risen from the present seven to nine (Malta, Libya, Jordan, Israel, Tunisia, Algeria, Egypt, Morocco and Cyprus). For the majority of these countries, TRpc will be significantly lower than the (very) critical threshold of 500 m^3/year/person.

Climate change

Most of the attention devoted to the impacts of enhanced concentrations of greenhouse gases has focused on a very limited aspect, i.e. the increase in

Table 2.2. TRpc: projections

Country	Total renewable water resources (TR) (km³)	1998 Population (thousands)	1998 Total renewable water resources per capita (m³) (TRpc)	2025 projection Population (thousands)	PGR[1] (%)	Total renewable water resources per capita (m³) (TRpc)	ΔTRpc[2] (%)
Albania	55.8	3,445	16,197	4,295	24.67	12,992	−19,80
Algeria	14.27	30,175	473	47,322	56.83	302	−36.15
Cyprus*	0.90	742	1,212	950**	28.03	947	−21.86
Croatia	143	4,494	31,820	4,243	−5.59	33,703	5.92
Egypt	58.3	65,675	888	95,766	45.82	608	−31.44
France	198	58,733	3,371	60,393	2.83	3,279	−2.74
Greece	58.65	10,551	5,559	10,074	−4.52	5,822	4.73
Israel	2.2	5,883	374	7,977	35.60	276	−26.26
Italy	167	57,244	2,917	51,744	−9.61	3,227	10.64
Jordan	1.7	5,956	285	11,894	99.70	143	−49.85
Lebanon	5.6	3,194	1,753	4,424	38.51	1,266	−27.79
Libya	0.6	5,980	100	12,885	115.47	47	−53.43
Malta*	0.0155	367	42	418	13.90	37	−11.71
Morocco	30	28,012	1,071	39,925	42.53	751	−29.84
Portugal	69.6	9,798	7,103	9,438	−3.67	7,374	3.82
Slovenia	33	1,919	17,196	1,738	−9.43	18,987	10.42
Syria	53.7	15,335	3,502	26,303	71.52	2,042	−41.70
Spain	111.3	39,754	2,800	37,500	−5.67	2,968	6
Tunisia	4.12	9,497	434	13,524	42.40	305	−29.81
Turkey	234	63,763	3,700	85,791	34.55	2,728	−30.17

Source: data elaborated from World Resources Institute (1998–99).
* Data relating to Cyprus and Malta taken from AQUASTAT, FAO (1995).
** This figure taken from Population Action International (1997).
[1] Population Growth Rate; [2] ΔTRpc = variation of per capita water availability.

annual average atmospheric temperature. However, global warming would do more than add a few degrees to today's average temperatures. For instance, "some of the most severe impacts to society and natural ecosystems are likely to result not from changes in temperature, but from changes in precipitation, evapotranspiration, runoff, and soil moisture" (Gleick, 1993, p. 128).

Available models, which are still of doubtful validity for accurate forecasting of the magnitude and the timing of global changes, are unsuitable for predicting the local impacts. Nevertheless, a consensus is emerging, at least with regard to the broad expected regional impacts of global warming on freshwater supplies and resources management.

As far as the Mediterranean region, and arid and semi-arid areas in particular, are concerned, these expectations can be summarized as follows (Gleick, 1993; Meyer, 1996; Chiamenti *et al.*, 1999):

- Although precipitation may increase slightly as a result of global warming, such an increase will have little impact in the arid and semi-arid parts of the Mediterranean basin, and is likely to be countered by higher average evapo-transpiration rates.
- The already high seasonal variability of rainfall is expected to increase further; while winter precipitation is projected to increase slightly, summer precipitation may decrease.
- All Mediterranean countries are expected to experience an intensification of the hydrological cycle; spatial patchiness of climatic conditions and inter-annual variability of rainfall are expected to increase, as well as the likelihood and severity of droughts and/or floods.
- Coastal aquifers could be affected by increased saltwater intrusion as the sea level rises.
- Climate change may enhance the demand for freshwater, particularly for agriculture and direct human consumption: a decrease in summer precipitation, whilst having little impact on the annual total, may nevertheless have significant effects on plant growth though extension of the summer period of water stress; the efforts of agriculture to adjust to climate change may lead to increased demand for irrigation, especially for soils with low water retention capacity; however, any increased use of irrigation water would be in conflict with the growing per capita demand for domestic uses induced by warmer average and extreme temperatures.

In short, although it is still hard to predict the local impacts of global warming on water supply and demand, it is reasonable to believe that water management problems presently faced by many Mediterranean regions are unlikely to be reduced, and may be further complicated and exacerbated by climate change. Perhaps the greatest certainty about the impacts of climate change is that both demands and supplies will become more uncertain. Because of the increasing volatility, matching future supplies with demands will become more difficult, and this will reinforce the need for water policies able to increase

Table 2.3. Average annual precipitation and evapo-transpiration rates

Country	Precipitation (mm)	Evapo-transpiration (mm)	Annual internal renewable water flow (INT) (km^3)
France	816	490	180
Greece	849	492	47
Italy	983	428	167
Portugal	886	474	38
Spain	662	432	116

Source: European Environmental Agency (1998a).

the adaptation capacity of resources management systems to droughts and chronic shortages.

2.3. Freshwater availability and groundwater resources management: southern EU Member States

2.3.1. *The spatial and temporal variability of freshwater supplies*

As showed in the previous paragraph, in global terms, southern EU Member States benefit from a relative abundance of renewable freshwater resources. Potentially, these countries have sufficient resources to meet national requirements, given the average rates of replenishments of their resources (see Table 2.3). Total estimated abstractions range between 9% (Greece) and 34% (Italy) of the total annual internal water flow (see Table 2.1).

However, although aggregated data based on long term average annual precipitation and evapo-transpiration rates tend to provide a reassuring picture, they hide significant geographical imbalances. Moreover, southern European countries, and in particular, arid and semi-arid Mediterranean areas, are characterized by a high seasonal and interannual variability of precipitation rates.

The geographical distribution of rainfall and evapotranspiration rates can give a broad idea of the spatial variability of regional water supplies. In Spain, precipitation rates (P) exhibit a significant geographical variability, ranging from 1315 mm in northern Spain to 380 mm in the Segura Basin; in some areas precipitation does not exceed 200 mm (Estrela *et al.*, 1996). Many regions exhibit arid or semi-arid characteristics, with average evapo-transpiration exceeding annual rainfall (see Table 2.4).

In Portugal (Table 2.5), there is an imbalance between the northern and southern basins ($P = 1800$ mm or even more in the north, P values around 700 mm or even less in the south). There are five regions with semi-arid characteristics or facing water scarcity problems: Alto Douro, Sul Tejo, Guadiana, Sado e Mira and Algarve (Estrela *et al.*, 1996).

Italy also faces a significant spatial and seasonal variability in water supplies.

Table 2.4. Spain: annual rainfall and evapo-transpiration in arid and semi-arid regions

Region	Annual rainfall (mm)	Annual evapo-transpiration (mm)
Guadiana	557	933
Guadalquivir	617	951
Sur	539	985
Segura	381	912
Jucar	520	802

Source: Estrela *et al.* (1996).

Table 2.5. Portugal: annual rainfall and evapo-transpiration in arid and semi-arid regions

Region	Annual rainfall (mm)	Annual evapo-transpiration (mm)
Alto Douro	683	1293
Sul Tejo	687	1381
Guadiana	564	1304
Sado e Mira	659	1327
Algarve	653	1689

Source: Estrela *et al.* (1996).

The northern regions, thanks to the Alps and to the natural storage capacity provided by glaciers and lakes, enjoy regular and abundant per capita endowment. In central and southern Italy, available resources are much lower, seasonal variability of runoffs is at the highest; the National Hydrographic Service (Servizio Idrografico Nazionale) has issued a map showing that a large part of the south suffers from consecutive periods of 100–150 days without rain (Massarutto, 1999). There are also areas characterized by structural water deficits. For example, on the island of Sardinia, the average annual precipitation is 753 mm and the mean annual potential evapotranspiration is about 1500 mm (Estrela *et al.*, 1996).

In Greece, the mean annual $P = 849$ mm. Two hydrological departments, Attiki and the Aegean Islands, are affected by severe scarcity problems as a result of unfavourable hydrogeology, low precipitation and demographic pressure. The Aegean Island region is the driest in Greece, with an average $P =$ 500 mm and a potential evapo-transpiration of 1250 mm. There are also scarcity problems in eastern Peloponnisos and in some areas of central Macedonia (Estrela *et al.*, 1996) (Table 2.6).

Even France, by far the richest country in terms of average internal renewable water flow (180 km^3/year), shows significant differences in the spatial distribution of water supplies, and Mediterranean regions, for example the south of Provence, experience significant seasonal variability in rainfall.

Table 2.6. Greece: annual rainfall and evapo-transpiration in arid and semi-arid regions

Region	Annual rainfall (mm)	Annual evapo-transpiration (mm)
Aegean Islands	500	1250
Attiki	900	1250
Northern Peloponnisos	800	1250
Eastern Peloponnisos	600	1250
Central Macedonia	600	1250
Kriti	900	1250

Source: Estrela *et al.* (1996).

2.3.2. *Groundwater resources and abstractions*

Southern European Member States benefit from a relative global abundance of groundwater resources and recharge rates. In France, there are three types of groundwater regions: 30% of these regions are situated in porous media; 10% in karst media and about 60% in other media (Koreimann *et al.*, 1996). The total average annual groundwater recharge is estimated at 100 km^3/year (World Resources Institute, 1997). In Greece, the groundwater potential in is around 10.3 km^3/year, while 7.4 km^3/year is karst groundwater (Koreimann *et al.*, 1996). In Italy, more than 50% of groundwater resources are in porous media extending over an area of about 158 000 km^2. There are aquifers situated in karst media and there are also smaller groundwater resources in volcanic rock media with an area of 13 488.78 km^2 (Koreimann *et al.*, 1996). The total average groundwater recharge is estimated at 30 km^3/year (World Resources Institute, 1997). In Portugal, the main aquifer systems are in porous media and karst: 29.4% of the national territory is an area of porous media while karst groundwater covers an area of 5500 km^2 (i.e. 6.2%). The principal sedimentary aquifer systems are located in littoral zones. The continental area of the country, corresponding to the main area of Portugal, is covered by igneous and metamorphic rocks. In this zone, the aquifers have low productivity but they are very important in satisfying several purposes locally (France *et al.*, 1996). The total annual groundwater recharge is estimated at 5.1 km^3/year (World Resources Institute, 1997). In Spain, more than one-third of the national territory contains groundwater aquifers: 16% of the whole country contains groundwater in porous media, 11% of the whole country consists of karst groundwater and other groundwater resources can be found in an area of 38 644 km^2 (8% of the whole country). According to MOPTMA (1998), groundwater recharge is estimated to be about 29 km^3/year; in terms of percentage, it represents about 26% of the total groundwater contribution.

In global terms, total abstractions are lower than the average annual recharge of national aquifers in southern European Member States. However, statistics at the national level mask local over-exploitation problems: for instance, in

Table 2.7. Water abstractions by source (%)

	Year	Surface waters	Groundwater
France	1995	85	15
Greece	1980	69	31
Italy	1985	77	23
Portugal	1990	58	42
Spain	1995	84	16

Source: European Environmental Agency (1998a).

Table 2.8. Use of groundwater per sector (%)

	Year	Public water supplies	Agriculture	Industry
France	1994	56	17	27
Greece	1990	37	58	5
Italy	1990	39	57.5	3.5
Portugal*	1990	38.6	22.8	38.6
Spain	1995	18	80	2

Source: Margat (1999); * World Resources Institute (1997).

Table 2.9. Water abstractions by source: breakdown per sector (% of total abstractions covered by groundwater)

	Year	Public water supplies	Agriculture	Industry
France	1994	57	20	40
Greece	1990	64	20	71
Italy	1990	91	29	7
Portugal		–	–	–
Spain	1995	22	18	5

Source: Margat (1999).

many areas, intensive abstractions have produced a significant lowering of the water table.[1]

Among southern European Member States, the largest consumer of groundwater is Italy (10.4 km^3/year), followed by France and Spain (6.0 and 5.53 km^3/year, respectively. The relative contribution of groundwater to total water supplies ranges between 42% (Portugal) and 15% (France) (see Table 2.7). With the exception of France and Portugal, agriculture is the most important user of groundwater (see Table 2.8). However, with the exception of Spain, the sector which is most dependent on groundwater supplies is the domestic one (see Table 2.9).

Again, as with resource endowments, national statistics hide regional differences in sector apportionment and the relative contribution of groundwater resources to total water supplies. In Italy, where aquifers account for roughly 90% of total household supply, northern regions are more dependent on groundwater than southern ones, where 15–25% of public supplies are obtained from surface waters, reservoirs and transfers; in the north and in the centre, agricultural abstractions basically occur in mountain and hilly areas where irrigation is often practised during the winter in order to prevent damage from hard frost, whilst in the south and the islands, groundwater is used intensively during the summer, both in internal areas or along the coastal plains (Massarutto, 1999). In Greece, in arid areas, agriculture almost exclusively relies upon groundwater (Barraqué, 1995). In Portugal, the Algarve region is totally groundwater dependent, with high levels of abstractions, especially in the summer period (Koreimann *et al.*, 1996). In Spain, where 12 million inhabitants (31% of the whole population) are supplied by groundwater, the population supplied with groundwater in the Jucar y del Sur river basin is about 50%, and groundwater is the only source in the archipelagos (Lopez-Camacho, 1996).

2.3.3. *Groundwater quality*

All over Europe, groundwater is endangered in numerous ways (European Environmental Agency, 1998b). Pollutants include heavy metals, chlorinated hydrocarbons, mineral oils (mainly attributable to leaching of dumping sites and municipal or industrial sources) and salts, nitrates and pesticides accumulated on farmland and dispersed through natural processes or farming practices. Saltwater is another potential pollutant in coastal regions where groundwater abstractions exceed the recharge rates of aquifers.

The geographical distribution and the spatial significance of groundwater pollution phenomena are difficult to assess because of the inhomogeneity of collected data and the lack of a uniform strategy for groundwater quality monitoring all over Europe (Vogel and Grath, 1998). For instance, countries apply different monitoring strategies and methods, e.g. monitoring might concentrate on areas foreseen or used as drinking water resources, and some pollutants are not found simply because they are not looked for.

Nevertheless, various reports based on the best available information have indicated salinization and contamination by nitrates and pesticides as the most typical or significant contamination problems affecting southern European regions (Estrela *et al.*, 1996).[2]

Groundwater salinization occurs via downward movement of salt accumulated on soils and seawater intrusion in coastal aquifers due to the lowering of the water table. Salinization due to downward movement of salt from leaching water is a relatively widespread phenomenon in areas with warm and dry climates such as those which characterize the Mediterranean basin. As far as

European countries are concerned, the most significant and widespread salinization problems can be observed in arid and semi-arid areas subjected to strong irrigation. Seawater intrusion due to overexploitation of coastal aquifers, which in southern Europe commonly arises from excessive abstraction for irrigation, is one of the major threats to sustainable use of groundwater (Estrela *et al.*, 1996).

In Spain, severe saline intrusion problems have been detected in the Segura and Jucar catchment areas. In Portugal, these problems arise in the central Algarve coastal region and the eastern Algarve coastal region. In Greece, seawater intrusion problems have been detected in 10 hydrologic departments, particularly in coastal areas and in the Aegean islands (Estrela *et al.*, 1996). In Italy, seawater intrusion has been detected in Sardinia and some parts of the Murgia, and in the Salento area.

Although local pollution due to municipal or industrial sources may be important, excessive nitrogen surpluses removed from farmland is, by far the main source of groundwater contamination by nitrates. According to model calculations, nitrate pollution in southern Europe tends to take place on a smaller scale than in other European areas. However, there are many areas where nitrate concentrations have significantly increased and exceed the maximum admissible concentration for water intended for human consumption laid down in the EEC Drinking Water Directive (80/778).

In Italy, the main problems are concentrated in the Po Valley. In Portugal, the areas most affected by nitrate pollution are located in Algarve, in the Tegus Valley and Alentejon. In Greece, high pollution loads have been detected in Attiki and Tessaly. In Spain, nitrate concentration has gradually increased in many areas where intensive farming is practised (see Table 2.10). In France, increasing nitrate pollution has been detected in Bretagne, Charentes and certain areas of Beauce (Barraqué, 1995).

Athough mineral fertilizers are the major artificial source of nitrogen to farmland, there are regions where, because of high livestock densities, manure is a major source of nitrogen on agricultural land. For example, levels of nitrogen inputs in Galicia are four times higher than the Spanish national average, in Bretagne, three times higher than the French average, in Lombardy, close to three times higher than the Italian average (European Commission, 1999).

Pesticides are important pollutants of Europe's groundwater. Many active ingredients are known to migrate easily into aquifers and, because of their toxic impacts, they may impede the use of groundwater for human consumption or make it more costly. The risk of pollution by pesticides is generally assessed according to groundwater quality data collected from various types of sampling sites, and on available information on the use of individual active ingredients. However, only a few pesticide measurements in groundwater are available for a restricted number of constituents: many active ingredients are not found in groundwater simply because they are not looked for. Moreover, only few

Table 2.10. Spain: nitrate contamination in groundwater (1995)

Catchment	Affected aquifers, maximum concentration values (mg/l NO_3)
Norte	Unknown
Duero	Esla-Valderaduey (50–150 in some areas); Region de los Arenales (50–190 in several wells); Segovia (90–250)
Tajo	Madrid-Talavera, La Alcarria, Ocana, Tietar (50 common; 100–150)
Guadiana	Mancha Occidental, Campo de Montiel (more than 100)
Guadalquivir	Depresion de Granada, Altiplanos de Ecija (160–295); Aluvial del Guadalquivir (Sevilla) (up to 147); Sevilla-Carmona (up to 290); Puerto de Santa Maria, Vejer-Barbate and Rota-Sanlucar-Chipiona (260–750)
Sur	Carchona-Castell de Ferro (150–660); Rio Verde (50–100)
Segura	Vega Alta del Segura, Valle del Guadalentin (50–100); Campo de Cartagena (100–240); Aguilas (340 at one point)
Jucar	Plana de Castellon, Plana Sagunto, Oropesa-Torreblanca, Plana de Valencia, Gandia-Denia (140–280); Mancha Oriental (120)
Ebro	Aluvial de Vitoria, Aluviales del Ebro (50–125); Aluvial del Gallego, Piedra Gallocanta (50–125)
C.I.Cataluna	Maresme-Llobregat area (greater than 500)
Baleares	Llana de Inca-Sa Pobla (60–250); Llucmajor-Campos (50–100); Sant Miquel Costa Norte (50–100)
Canarias	300 in widespread areas

Source: Candela and Varela (1998).

Member States carry out regular direct surveys on use of pesticides by farmers based on a representative sample of agricultural holdings (Sweden, Netherlands, UK) (European Commission, 1999).

Use of pesticides differs from country to country. An analysis of the regional distribution of dominating substances shows that in France, Italy, Portugal, Spain and Greece, fungicides dominate. Between 1991 and 1995, there was a downward trend in the sales of pesticides measured by weight of active ingredients (European Commission, 1999) in Member States, although this does not necessarily indicate a decrease in environmental impact since the range of pesticides in use has changed (European Environmental Agency, 1998b). The declining trend was reversed in 1996, when total sales in the EU rose by 6%.

Generally speaking, the EU Member States which have experienced the greatest reduction of pesticides sales are those countries characterized by a drastic cut in agricultural prices (Finland, following EU accession; −46% of sales between 1991 and 1996) or which have adopted specific policies to reduce pesticide use (Finland; the Netherlands, −43%; Austria, −21%; Denmark, −21%; Sweden, −17%) (European Commission, 1999).

Over the period 1991–1996, in France, Italy and Spain, respectively the first (31% of total sales in the EU in 1996), second (16%) and fifth (11%) most important markets for pesticides, sales of pesticides decreased by 11%, 17% and 15%, respectively. However, in 1996, sales in Spain and France were up

+19% and +11%, respectively, compared to 1995 (European Commission, 1999).

2.3.4. *Groundwater resources management: national legislation and EC policies*

In Member States, there has been an increase in the regulatory power of public authorities in respect of water use: in general terms, increasing state regulation and centralized water management has been a common feature in southern European countries, while involvement of local authorities and subsidiarity have remained a typical feature of the institutional set-up in Northern countries. The share of public inland waters has tended to increase, and abstractions and wastewater disposal has been increasingly subject to licensing and mandatory regulation. However, groundwater was only later involved in this trend, because it was hidden, poorly known, considered as a local and minor resource, and above all often left to the owners of the land above (Barraqué, 1998).

In France, until the 1992 draft water law, groundwater was generally considered to be private, with the exception of over-exploited areas where uncompensated State demanialization would occur. Because of this private appropriation tradition, there is little groundwater management administration, and even a certain lack of knowledge of its availability (Barraqué, 1998). Even in Italy, where inland waters have progressively entered the public domain during the course of the last century, until very recently, groundwater abstraction was free and considered to be part of the rights of landowners, whilst exploitation of surface waters usually required an abstraction and use licence from the competent authority. It was only in 1994 that this dual system ended: the law 36/1994 states that all water usage, including groundwater abstractions, need to be licensed. However, implementation of this legal provision is not easy: some tens of thousands of private abstractions need to be documented and monitored (Massarutto, 1999).

In Spain, which has a long tradition of centralized water policy and demanialization of surface water, it was only in 1985 that groundwater was placed in the public domain. The 1985 water law gave farmers the option to choose between keeping the former ownership system and abandoning their rights, with State compensation taking on the form of protection of their abstraction rights for 50 years. According to Barraqué (1998), "most farmers refused to declare their rights, and among those who did, few accepted this sort of demanialisation" (p. 85).

In Portugal, groundwater was traditionally considered to be part of the land ownership rights. A licensing system for groundwater abstractions was enforced only in 1977 (decree law 376/1977), in order to regulate groundwater usage in Algarve and other southern areas almost exclusively dependent on aquifers. The licensing system was then generalized in 1994 to the entire country (Barraqué, 1998).

As far as groundwater quality management is concerned, Member States' regulations have been fostered by EC policy initiatives directly or incidentally addressing groundwater pollution problems. The Drinking Water Directive (80/778/EEC), although it did not specifically address groundwater, laid down maximum admissible concentrations of nitrates (50 mg NO_3/l; guide level 25 mg NO_3/l) and pesticides (0.1 μg/l for each active ingredient and 0.5 μg/l for the total pesticide) for water intended for human consumption.

Specific EC groundwater legislation started in 1980 with the Directive 80/68/EEC aimed at preventing all direct discharges of various dangerous substances into groundwater. Activities such as farming which are likely to lead to indirect discharges were also addressed by the Directive, which included pesticides and nitrates in the list of dangerous substances. However, Member States have had considerable problems in incorporating the directive into national legislation (Krämer, 1990).

Ten years later came the Nitrate Directive (91/676/EEC), aimed at reducing or preventing water pollution from the application and storage of inorganic fertilizer and manure on farmland. Although the Nitrate Directive was not directed towards groundwater in particular, but towards eutrification problems in rivers and estuaries, it has fostered the implementation of specific programmes to reduce the level of contamination of aquifers (Barraqué, 1998). In particular, Member States were required to identify nitrate vulnerable zones and design and implement action programmes by December 1993. However, as was highlighted in a report recently released by the EU Commission (1997) (COM(97)473) generally speaking, there has been a significant lack of progress in Member States' implementation of the Directive. For instance, 4 years after the deadline, only four States (Denmark, France, Luxembourg and Spain) have brought into force laws, regulations, and administrative provisions necessary to comply with the Nitrate Directive. In Italy, for example, the Nitrate Directive was implemented only in 1999 (decree law 156), and only a preliminary set of vulnerable zones where the Directive envisaged that Codes of Good Agricultural Practice should be applied have already been identified.

The European Commission (1996, 1997) has recently released a proposal for an Action Programme for Integrated Groundwater Protection and Management (COM(96)315 Final) and a proposal for a Council Directive establishing a Framework for Community Action in the Field of Water Policy (COM(97)49 Final),[3] which are intended to form the framework for the whole EU water policy.

As far as the former document is concerned, the general objective set up in the Commission's proposal is "to ensure protection and use of groundwater through integrated planning and sustainable management aiming at preventing further pollution, maintaining the quality of unpolluted groundwater, restoring, where appropriate, polluted groundwater, as well as preventing the over-exploitation of groundwater resources" (European Commission, 1996). The main lines of the action programme are contained in the Annex of the Commission's proposal (1996), and include the following general planning principles and regulatory strategies:

- A sustainable quality management should protect and preserve all groundwaters ... Actions to achieve this should be based on the principles of prevention, action at source, and that the polluter should pay.
- Within the overall objective of protection of groundwater, relieving the environmental pressure from diffuse sources should have the highest priority because the largest quantities of groundwater are found and formed in the countryside.
- All possibility and strategies to lessen the impact of diffuse sources such as nitrate and plant protection products should be explored. Introduction of economic instruments amongst other measures should be included. These instruments could be based on further incentives encouraging environmentally friendly sustainable farming. Use of the principle of internalizing the environmental costs with the help of taxes and levies directly aiming at the consumption of chemical fertilizers and plant protection products, excessive application of manure from intensive livestock farming, etc. could be explored.
- The development of codes of good agricultural practice for environmentally compatible production should be at the centre of actions taken. Appropriate measures to monitor compliance with the codes of good agricultural practice should be established. As compliance with the codes in itself may not be sufficient to achieve the objectives in certain regions, measures of a further-going nature to ensure environmentally compatible production could be developed. Possibilities for using the principle of cross-compliance should be explored in this context. In order to avoid distortion of competition and to create so called win–win situations benefiting both the environment and the farmers, strategies to compensate farmers should be developed also.

2.3.5. *Groundwater monitoring systems*

While the monitoring of groundwater quantity has a relatively long tradition in Europe, with the oldest networks being in operation since 1845, monitoring of groundwater quality on a regular basis has been undertaken only quite recently in Member States.

Despite the increasing efforts undertaken in Member States to improve their monitoring capacity, "groundwater continues to be a largely misunderstood ... resource" (Hernández-Mora *et al.*, Chapter 5, this volume, p. 108). For instance, some southern European countries, namely France, Greece and Italy, lack a homogeneous nation-wide monitoring system for groundwater quality and quality. Moreover, as Koreimann *et al.* (1996) emphasize, the connection between monitoring activities and legal obligations is surprisingly low: only some of the Member States monitor groundwater in their assessment of national legislation and only Portugal monitors due to EC legislation.

The need to improve Member States' monitoring capacity in order to acquire the information required to implement more effective and efficient water policies has been stressed in both the Commission's preparatory acts mentioned in the previous paragraph.

In the proposal for the Action Programme for Integrated Groundwater Protection and Management (Annex) (European Commission, 1996) it is stated that improvement in monitoring capacity should be achieved through the establishment of national monitoring programmes, aiming, *inter alia*, at identifying areas with groundwater of importance for present and future drinking supply and for particular ecological functions, and areas where groundwater is particularly sensitive to pollution as a consequence of particular geological or climatic conditions, the nature of the soil or man-made influences.

According to the second preparatory act (European Commission, 1997), which is concerned with the establishing of a Framework for Community Action in the Field of Water Policy (Technical Annex V), Member States shall ensure the establishment of programmes for the monitoring of water status in order to establish a coherent and comprehensive overview of water status within each River Basin District. Such programmes shall cover monitoring of the chemical and quantitative status of water, including groundwater, and they ought to be operational by 31 December 2001.

Notes

1. Examples are Campo de Dalia (near Almeria) and Sierra de Crevillente (Alicante), in Spain; many areas located in the Po valley, Sicily (near the city of Augusta) and Sardinia (Iglesias), in Italy; the plain of Argolide in Greece (Margat, 1999); the region of Algarve in Portugal.
2. These pollution phenomena and their interlinkages with agricultural activities are illustrated in Chapter 3.
3. The proposal has been subsequently amended in 1998 (COM(1997) 614, COM (1998) 76, COM (1999) 271).

References

Barraqué, B., 1995, Les Politiques de l'Eau en Europe, Paris: Editions La Decouverte.

Barraqué, B., 1998, Groundwater management in Europe, in Hilding-Rydevik, T., Johansson, I., eds., How to Cope with Degrading Groundwater Quality in Europe, Stockholm, Swedish Council for Planning and Coordination of Research, pp. 74–101.

Brouwer, F., Falkenmark, M., 1989, Climate-induced water availability changes in Europe, Environmental Monitoring and Assessment, 13, 75–98.

Candela, L., Varela, M., 1998, Groundwater pollution from mineral fertilizers and pesticides in Spain, Proceedings of the workshop Agricultural Threats to Groundwater Quality, Zaragoza, 27–30 October 1996, UNESCO.

Chiamenti, E., Dosi, C., Ramieri, E., 1999, Climate change and the Mediterranean Basin: an overview and the impacts upon water resources and m,anagement, Paper presented at the 2nd SAGA Workshop, Fondazione Eni E. Mattei, Venice, 23–24 April, 1999.

Calow, R., 1998, Groundwater resources degradation: socio-economic impacts and their mitigation, in Hilding-Rydevik, T., Johansson, I., eds., How to Cope with Degrading Groundwater Quality in Europe. Stockholm: Swedish Council for Planning and Coordination of Research, pp. 38–56.
Estrela, T., Marcuello, C., Iglesias, A., 1996, Water Resources Problems in Southern Europe: An Overview Report, European Topic Centre on Inland Waters, European Environmental Agency.
European Commission DG XI, 1997, The Implementation of Council Directive 91/676/EEC Concerning the Protection of Waters against Pollution Caused by Nitrates from Agricultural Sources, Report COM(97)473.
European Commission, 1996, Proposal for an Action Programme for Integrated Groundwater Protection and Management, COM(96)315 Final.
European Commission, 1997, Proposal for a Council Directive Establishing a Framework for Community Action in the Field of Water Policy, COM(97)49 Final.
European Commission, 1999, Agriculture, Environment, Rural Development: Facts and Figures – A Challenge for Agriculture, http://europa.eu.int/comm/dg06, July 1999.
European Environment Agency, 1998a, Statistical Compendium for the Second Assessment. Office for Official Publications of the European Communities, Luxembourg.
European Environmental Agency, 1998b, Europe's Environment: The Second Assessment. Oxford: Elsevier Science Ltd.
Eurostat, 1995, Europe's Environment: Statistical Compendium for the Dobris Assessment, Office for Official Publications of the European Communities, Luxembourg.
FAO, 1995, AQUASTAT. http://www.fao.org/waicent.
France, S.J., Marcuello, C., Ashley, S.J., Nixon, S.C., Lack, T., 1996, International Water Databases, Topic Report 16/96, European Topic Centre on Inland Waters, European Environmental Agency.
Gleick, P.H., 1993, Effects of climate change on shared fresh water resources, in Mintzer, I.M., ed., Confronting Climate Change. Risks, Implications and Responses, Cambridge: Cambridge University Press, pp. 127–140.
Koreimann, C., Grath, J., Winkler, G., Nagy, W., Vogel, W.R., 1996, Groundwater monitoring in Europe, European Topic Centre on Inland Waters, European Environmental Agency.
Krämer, L., 1990, EEC Treaty and Environmental Protection, London: Sweet and Maxwell.
Lopez-Camacho, B., 1996, La gestion de las Aguas Subterraneas en los Abastecimientos y en los Regadios, Conflictos, Soluciones y Tendencias, Paper presented at the Curso de Verano de la Fundacion de la Universidad Autonoma de Madrid y el Foro del Agua, July 1996. http: www.pangea.org/org/foroagua.
Lundqvist, J., Gleick, P., 1997, Comprehensive Assessment of the Freshwater Resources of the World, Stockholm Environment Institute.
Margat, J., 1999, Utilisations et Abuses des Eaux Souterraines Dans les Pays Méditerranèens, Paper presented at the Cursos de Verano de la Universidad Complutense, Fundacion Marcelino Botin, Madrid, 2–6 August, 1999.
Massarutto, A., 1999, Agriculture, Water Resources and Water Policies in Italy, Working Papers (33.99), Fondazione ENI E. Mattei, Milano.
Meyer, B.W., 1996, Human Impact on the Earth. Cambridge: Cambridge University Press.
Ministerio de Obras Públicas, Transporte y medio Ambiente (MOPTMA), 1998, Libro Blanco del Agua en Espana. Documento de Sintesis. Centro de Publicaciones, MOPTMA.
OECD, 1997, OECD Environmental Data Compendium 1997, Paris: OECD.
Population Action International, 1997, Sustaining Water, Easing Scarcity: A Second Update, http://www.populationaction.org/why_pop/water/water-toc.htm.
RIVM/RIZA, 1991, Sustainable Use of Groundwater. Problems and Threats in the European Communities, Report No. 600025002. Ministerseminar Groundwater, Den Haag, 26–27 November 1991.
Vogel, W.R., Grath, J., 1998, Groundwater in Europe: state and trends in quality and quantity, in Hilding-Rydevik, T., Johansson, I., eds., How to Cope with Degrading Groundwater Quality in Europe, Stockholm: Swedish Council for Planning and Coordination of Research, pp. 11–23.
World Resources Institute, 1997, World Resources 1996–97, New York: Oxford University Press.
World Resources Institute, 1999, 1998–99 World Resource. A Guide to the Global Environment, New York: Oxford University Press.

CHAPTER 3

Agricultural Impacts on Groundwater: Processes, Modelling and Decision Support

Andrea Giacomelli, Carlo Giupponi and Claudio Paniconi*

3.1 Introduction

Agriculture, like other economic activities, can have negative impacts on the environment and on groundwater in particular, with consequences for human health as well as the environment (FAO, 1979, 1996). On the one hand, the use of groundwater by agriculture can cause quantitative resource depletion and qualitative deterioration due to overuse or misuse, such as encroaching saltwater and salinization in coastal areas. On the other hand, the recharge of aquifers with water leached out from cultivated fields can lead to water pollution by agrochemicals (fertilizers and pesticides). Of the many types of compounds that can contaminate soils and groundwaters, those most closely associated with agricultural practices, and whose impacts will be discussed in depth in this contribution, are nutrients, pesticides and salt.

Once contaminants enter the subsurface, they are subject to complex physical, biological and chemical processes that transport and transform them. Accurate monitoring, together with mathematical models capable of realistically representing these processes, can be useful tools for studying the behaviour and effects of contaminants in subsurface waters. The models can be used for predicting future migration and for assessing alternative remediation strategies, and can provide information useful to the decision maker responsible for monitoring and planning resource utilization and for devising improved water resource management practices. An understanding of the fundamental processes that control the fate and transport of subsurface pollutants is critical to the development of simulation models for the prediction and analysis of these phenomena. The most comprehensive models are those constructed using so-called process- or physically based approaches, although in many cases and for various reasons, more empirical or conceptual models are used as well.

There is a wealth of concepts and issues connected to the modelling of natural phenomena and to the effective use (or, just as commonly, subtle abuse) of models. These range from the difficulties in handling inherent variability,

C. Dosi (ed.), Agricultural Use of Groundwater, 35–75.

scale, and nonlinearity to the need for effective pre- and post-processing of model inputs and outputs, transforming simulation results to useful information. We dwell in particular on the latter issue, which we feel:

- goes beyond mathematical-physical aspects to touch on current and emerging issues related to the practical application and integration of models within agricultural and environmental policy
- is relevant to general efforts to bridge the gaps between the study of natural phenomena and the analysis and implementation of socio-economic policies
- can contribute to the definition of mutually compatible agricultural and water management policies
- brings together field work and experimental activities with simulation models and advances in information technology such as geographic information systems (GIS) and remote sensing – all essential elements in the evolution of decision support systems that aim to contribute to more effective environmental monitoring and resource management.

This contribution deals in general with the assessment of agricultural production processes in relation to the environment, with specific emphasis given to groundwater resources in southern European countries and the Mediterranean. We will describe these production processes, their relation to and impact on groundwater, and the basic mechanisms and dynamics involved in the flow of water and transport of contaminants in soil and aquifer systems. An example of different modelling approaches is given, and the limitations and advantages of various approaches are discussed.

3.2. General concepts

3.2.1. *Pollution from agricultural sources*

Agricultural activities always have significant effects on the environment; some of these effects are positive, others are negative. Among the latter are the phenomena of degradation of water resources identified as pollution from agricultural sources, PFAS (EC, 1991). PFAS can be defined as a series of possible negative changes to the state of environmental variables (qualitative or quantitative) due to the introduction of substances or techniques used in primary production in the agro-ecosystem. Typical examples are the appearance of herbicides in groundwater as a consequence of weed control, or the rise in nutrient concentrations in runoff waters caused by fertilization. In general, the main cause of these phenomena is that the efficiency of agricultural production processes never reaches 100% and, more importantly, tends to decrease with the intensity of agricultural production systems.

Apart from those pollution phenomena deriving from intensive livestock rearing that produce wastes (liquid manure in particular) that could be

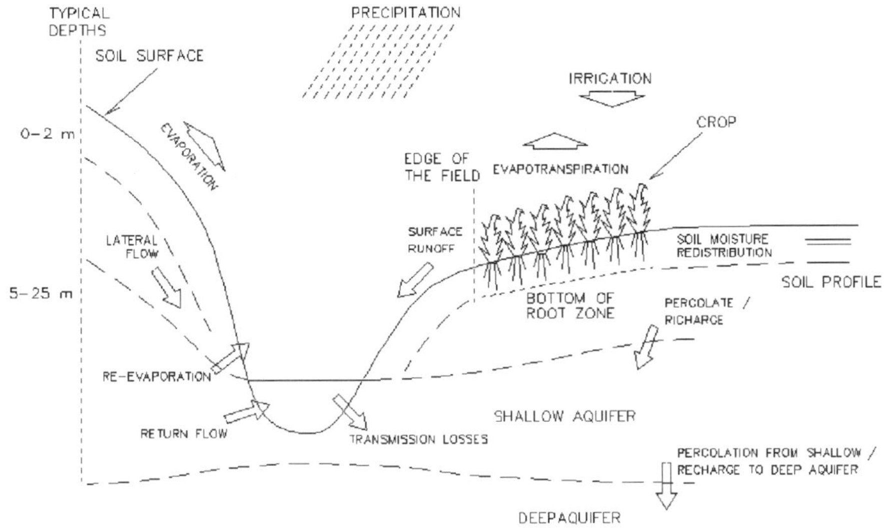

Figure 3.1. The hydrological cycle in agro-ecosystems.

described as point sources, the emissions of pollutants from agricultural fields are in general defined as diffuse (or non-point) sources (Novotny and Olem, 1994). These are characterized by the fact that they introduce polluting substances into the environment over wide areas in a spatially distributed way (i.e. with variations depending on the location), at irregular temporal intervals and under the effects of interactions between anthropic and natural variables. The hydrologic cycle is the main engine of these phenomena (Figure 3.1) and water is generally both the vector of pollutants and the receptor of impacts.

Two main categories of pollutants should be dealt with in examining the pollution of groundwater from agricultural sources: nutrients and pesticides. Nutrients are mineral or organic compounds containing mainly nitrogen, phosphorus or potassium that occur in soils and water through natural processes (animal manure, microbial breakdown of organic matter) and through human activities or human-controlled land use practices (application of fertilizers, sewage releases, soil cultivation, livestock production). In dissolved form, the principal nutrient-derived pollutants are nitrate (NO_3^-), ammonium (NH_4^+), and phosphate (PO_4^-). Nutrients pose risks to both human health (e.g. blue baby syndrome caused by drinking water rich in nitrates) and aquatic ecosystems (e.g. eutrophication of surface waters due to phosphates). Pesticides are synthetic organic compounds used to control weeds, insects, and other organisms, and their presence in groundwater is exclusively anthropogenic. At high concentrations in drinking water, pesticides can pose serious risks to human health; similar risks are posed for non-mammal biota (e.g. fish) in freshwaters.

Releases of fertilizers and pesticides can be referred to as a direct impact of agriculture on groundwater, in the sense that the contaminants are directly introduced into the environment as agricultural inputs.

3.2.2. *Salinization and seawater intrusion*

Salt is found naturally in seas and in geological formations (salt domes in sedimentary basins, for example), and can be introduced artificially as a by-product of fertilizers or land-disposed wastes. Even the digging of wells can in some cases destroy natural barriers (e.g. impermeable layers) and put freshwater in communication with salt water (Atkinson, 1987). Because it is so widespread, salt constitutes a particularly important category of groundwater pollutant, occurring in groundwater and soils via seawater intrusion or other salinization processes. Saltwater is not considered a health risk per se, but its presence in soils or underground waters used for industrial, agricultural, or domestic purposes can have grave economic consequences. In coastal aquifers, prolonged overpumping of groundwater can lead to an encroachment of the interface between seawater and freshwater, through intrusion and/or upconing (Sherif and Singh, 1996; Bear *et al.*, 1999).

If the aquifer is overexploited for irrigation water, then the resulting intrusion or upconing is referred to as an indirect impact of agriculture on groundwater, in the sense that the contamination is not directly introduced as an agricultural input (as with nutrients or pesticides), but rather is induced by extracting too much water to satisfy irrigation needs.

3.2.3. *Resource depletion and sustainable agriculture*

Considering the conflicts between agriculture and the environment (soil loss by erosion, water pollution, salinization, etc.), especially in relatively dry areas such as southern Europe and the Mediterranean, many authors treat these phenomena within the broader concept of desertification. Desertification can, in fact, be considered as a complex phenomenon determining reductions of biological and economic productivity and increased pollution (Perez-Trejo, 1992). In this regard water is obviously a crucial natural resource, and the Mediterranean is one of the areas where such conflicts are particularly felt.

The intensity of agricultural impacts on groundwater is determined by a combination of abiotic (climate, geomorphology, etc.), biotic (vegetation and fauna), and merobiotic (soil) factors. When performing an environmental impact assessment of an agricultural system it is, therefore, important to consider the territorial context of implementation. The compatibility, or inversely the conflict, of a production system with the environment is a function of the interactions between its environmental pressures and the vulnerability of the land. For the purposes of the present work, it is useful to consider these aspects (impacts, compatibility, vulnerability, etc.) within the broader concept of sustainability. Following the FAO definition of sustainable agricultural development

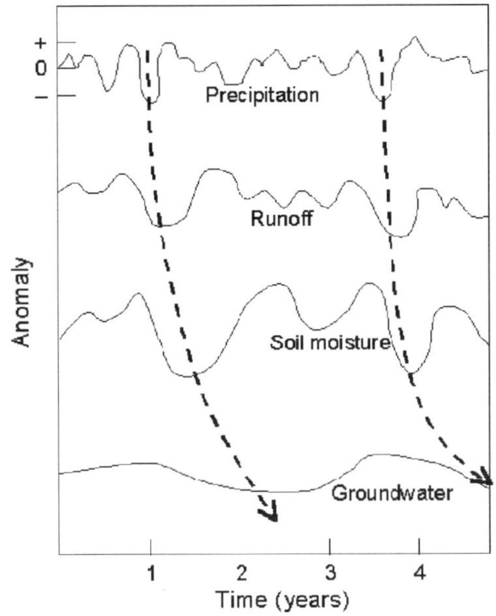

Figure 3.2. The subsurface as buffer and filter: propagation and attenuation of a rainfall signal on its downward journey through a porous medium (adapted from Entekhabi, 1995).

(FAO, 1996), the perspective adopted herein is to contribute to the identification of agricultural activities able to manage and conserve groundwater resources and to direct technological and institutional changes, ensuring the satisfaction of society's present and future water needs.

3.2.4. *Basic territorial units: hillslope, aquifer, and watershed*

The hillslope, the catchment (or watershed), and the aquifer are the most common territorial units used in studies dealing with the hydrological cycle (Figure 3.1) and associated processes such as transport of solutes in natural porous media. One of the most important distinctions to be made, applicable to all three units, is the distinction between the unsaturated zone, variously called the soil or vadose zone, where the pores contain air and water, and the saturated or groundwater zone, where only water is present. The two zones are separated by the water table. The water table rises and falls, and the degree of saturation in the vadose zone increases and decreases, in response to rainfall, evaporation, groundwater pumping, irrigation and other external forcing variables.

The nature of these fluctuations in response to natural atmospheric forcing, and the degree to which the subsurface can act as a buffer and a filter, is illustrated in Figure 3.2, where the propagation of a rainfall signal, which is highly variable, is dampened (and also shifted in time) as it passes through successive storage components associated with the surface (runoff), near-surface (soil moisture), water table, and deep subsurface (groundwater). An anomalous event (excessive rainfall leading to flooding, extended dry period leading to drought, a major contaminant spill on the surface) can be identified and monitored in different ways across these various layers of the subsurface as the dominant signal and surrounding noise are attenuated and smoothed out.

The hillslope is the smallest of the basic units, with a typical length of hundreds of meters and a depth in the order of meters. It is an ideal unit for conducting field trials and for obtaining ground measurements relating to processes such as percolation, seepage, flow through macropores, and the exchanges of water and energy between the soil, root zone, surface vegetation, and atmosphere. One- and two-dimensional models are often adequate for hillslope scale studies.

An aquifer system can contain a number of stratified aquifers separated by less permeable geological units such as aquitards and aquicludes. An individual aquifer can be unconfined (in contact with the unsaturated zone) or confined, and can range in scale from local units of modest depth and extent to regional scale aquifers deep below the surface of the Earth.

The watershed is the fundamental unit used in many branches of hydrological research such as studies of land–atmosphere interactions for climate change, flood frequency analysis for extreme events, and sediment transport and erosion in geomorphology. A watershed connects the atmosphere, land surface, subsurface, and streams, and is a conduit for the endless transformation and transportation of energy, water, solutes, and sediments. It constitutes a self-contained hydrological unit in the sense that it has a natural topographic boundary separating it from surrounding catchments and a natural outlet: a particle of water that falls on its surface cannot flow (overland) beyond this boundary, and any water that does not evaporate or percolate to deep aquifers eventually reaches the outlet, either via overland or subsurface routes. Increasingly, point and non-point source water pollution problems are being addressed at the watershed scale, along with more generic water management issues, as is made explicit in the recent Water Framework Directive of the European Commission (EC, 2000). The emergence of the catchment as a reference unit in water directives is aided by the widespread availability of digital terrain data and GIS-based topographic analysis software (Band, 1986; Jenson and Domingue, 1988), with which a watershed's boundaries can be delineated, its stream network identified, and its area subdivided into smaller and still hydrologically distinct units, from subcatchments right down to hillslopes. Watersheds can range in size from a few hectares for the smallest subcatchments to millions of square kilometers for continental scale river basins.

With regard to subsurface contaminants from agricultural practices, saltwater is most appropriately addressed from an aquifer perspective since it most commonly arises from excessive pumping of groundwater, nutrients from a watershed focus because of their close interactions with plants and the vadose zone and their important impacts on local streams and lakes, and pesticides on a case-by-case basis given the complexity of these substances (they can variously be adsorbed onto soil grains, biodegraded in the shallow subsurface, or percolate into deep confined aquifers).

Beyond the hydrogeologically oriented hillslope, catchment, and watershed units, it is worth mentioning that alternative basic study units have been proposed and used, especially when there is a need to characterize hydrologic response with respect to some aggregates of topography, soil, land cover, land use, geology and climatic features, giving rise to what one may term ecologically similar units or hydrological response units (Moore *et al.*, 1993; Kite, 1995). Such classifications are especially amenable to GIS processing.

3.2.5. *Scale, variability, non-linearity, and other issues*

The parameters and processes that characterize natural systems such as watersheds, soils, and aquifers exhibit tremendous variability in space and time. Rainfall, streamflow, vegetation cover, topography, soil texture, hydraulic conductivity, and water table levels, to name a few, reflect and respond to climate, land use and tectonics, and the interactions between these dominant forces. There is variability both within a given system (e.g. the saturated hydraulic conductivity of a single hillslope can change by orders of magnitude from one point to another), and between systems (e.g. the magnitudes and characteristic time scales of outflow are very different between clayey, sandy and silty hillslopes).

Some hydrological processes exhibit characteristic length and time scales of centimeters/minutes, while others act over hundreds of kilometers or thousands of years. Still other processes have no specific correlation or length scale. For instance, the heterogeneity of the porosity of a laboratory soil column may be characterized by its grain size distribution; in small field plots, plant root and earthworm channels may be significant in deriving a measure of porosity; for regional aquifers geological fractures and faults become important in defining this quantity.

Non-linearity is a feature that is encountered in numerous functional dependencies that govern the behaviour and response of water and solutes in porous media. In an unsaturated soil, for example, the moisture content and hydraulic conductivity depend in a highly non-linear manner on the pressure head. The pressure head in the vadose zone is highest (0) when the soil is saturated and lowest (negative values representing suction) when the soil is very dry and its moisture content reaches residual values. Aside from complicating the mathematics and numerics of modelling subsurface systems, non-linearity also makes it important to characterize the sensitivity of a given physical process or model output to the components that determine it, as small changes in an underlying parameter or in a model's initial and boundary conditions can result in large

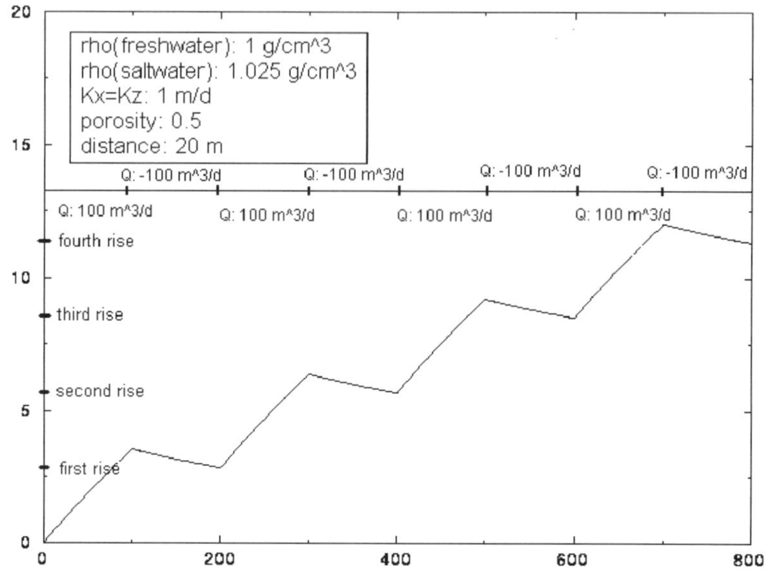

Figure 3.3. Hysteresis and irreversibility in groundwater exploitation: groundwater upconing is compounded via repeated cycles of pumping and active recovery, and the saltwater interface inexorably rises to reach the pumping well, situated at 20 m from the initial position of the interface (from Hassler *et al.*, 1999).

changes in an observation or output, and even unstable behaviour. This is one of the factors that wreaks uncertainty on model prediction.

Subsurface flow and transport dynamics often exhibit hysteresis or irreversibility, whereby a forward process (e.g. soil wetting or solute adsorption) follows a time path or curve that is different from the backward or reverse process (e.g. soil drying or solute desorption). Another example of irreversibility is related to saltwater upconing during groundwater exploitation in coastal aquifers. Upconing occurs when the saltwater–freshwater interface below a pumping well rises in response to pumping. Downconing is the reverse process, and can occur either actively (water is pumped back into the well to recharge the aquifer) or passively (extraction of water from the aquifer is halted and only natural recharge, determined by rainfall, takes place). Upon repeated cycles of pumping and active recovery, with the same pumping rate used for both extraction and injection of water, the saltwater–freshwater interface may not return to its initial position, as shown in Figure 3.3. This type of dynamics has practical implications for groundwater exploitation. If pumping is stopped to allow the aquifer to recover and the cone of saltwater to retreat, it is evident that the recovery time must be much longer than the pumping time, even when recovery is enhanced by injecting water back in at the same rate. Normally it will be necessary to begin pumping again from a given well after a certain recovery period (for instance, pumping during summer months when there is little rainfall and agricultural water needs are high, and passive or active recovery during fallow or winter months). Repetition of such a regime over extended periods of time will cause the interface to eventually reach the pumping well, causing costly saltwater contamination of drinking and irrigation water.

3.3. Groundwater pollution from agricultural sources

Three main phases can be distinguished in the phenomena of diffuse pollution from agricultural sources: generation, transport and discharge. The time and space dimensions are fundamental aspects of pollution from agricultural sources. In order to be understood and fully described, each phenomenon must therefore be investigated during all three phases. These phases are generally manifested in sequence and with a well-staggered spatial distribution.

3.3.1. *Generation of loads*

The phase of generation of the diffuse loads is when a polluting event takes place as a consequence of an agronomic practice applied to a cultivated field, for example the spraying of herbicides or spreading livestock manure. The spatial scale is, in this case, that of the field, i.e. a portion of land cultivated in the same way and homogeneous from the environmental viewpoint, characterized in particular by the same soil and climate. The time scale of the phenomena is widely variable: some occur as practically instantaneous impulses, such as the atmospheric drift of pesticides at time of application, others instead have much longer dynamics that may even last for years, such as nutrient releases in water following phenomena of mineralization and leaching in soils.

3.2.2. *Transport*

The transport phase of the diffuse loads is when a pollutant of agricultural origin moves across the environment; for example, when a certain amount of pesticide lost from a cultivated field percolates below the root layer through the unsaturated layer, before it reaches the groundwater. It is interesting to note that during the transport phase, the polluted environmental resources (air, water or soil) may at the same time be both vectors and receptors of the pollution; therefore, depending on the environmental compartment and territorial ambits to be considered, different successions of transport and discharge phases can sometimes be identified.

The spatial scale for the analysis of transport phenomena is generally much larger than the field, typically the area involved is the catchment (or watershed). The time scale of the phenomena is also variable in this case: some pollution events have dynamics in the order of hours (e.g. runoff of water containing agrochemicals after rainfall), but, especially when the problems are addressed in terms of whole ecosystems and on the scale of large territories, the phenomena show multi-annual dynamics (e.g. contamination of aquifers).

3.3.3. *Discharge*

The discharge phase is when the pollutant of agricultural sources reaches the resource that can be considered as the final receptor; for example, when the amount of pesticide lost from a cultivated field reaches an aquifer, the sea or the mouth of a river or lake. The spatial scale is generally very large as it is usually necessary to understand the effects of the polluting events in the entire water body (a whole lake, for example). The time scale for fully describing the phenomena is normally in the order of months or years.

3.4. Salt contamination of aquifers

Generally speaking, contamination by salt reduces water quality for human consumption, while agriculture and irrigation are extremely sensitive to the accumulation of salt in the soil zone. Saltwater is often in a delicate equilibrium with freshwater aquifers. When the equilibrium is disturbed, the transition to a new balance could, depending on the scale of the phenomena, take decades to achieve (e.g. in the case of disposal of minor salt-containing wastes), centuries (regional scale saltwater intrusion), or millennia (salt dissolution in deep formations) (Frind, 1982). A proper evaluation of the environmental impacts and economic effects associated with salt contamination phenomena therefore requires monitoring and analysis of the short, medium, and long-term response of the threatened system. It is important to adequately characterize the aquifer flow regime and its natural patterns of land recharge and sea discharge, for instance. For heavily utilized aquifers in semi-arid regions, the flow regime will be particularly sensitive to replenishment by irrigation, artificial recharge, and the rainfall that is effectively infiltrated and is not directly lost to surface runoff and evaporation.

Salt is non-reactive, and thus as a solute it is not as difficult to model as some other substances which undergo complex chemical and biological transformations in soils and aquifers. On the other hand, the presence of salt alters water density in such a way as to induce important effects on the pressure and flow fields, and these effects pose some mathematical and numerical difficulties of their own. Density-dependent phenomena in groundwater flow and transport are those in which differences in density between the components of the system have a strong influence on its evolution (Kolditz *et al.*, 1998). Typically, the contaminant will be driven downward by gravitational force through the unsaturated zone. If the density of the contaminant is higher than that of water, the

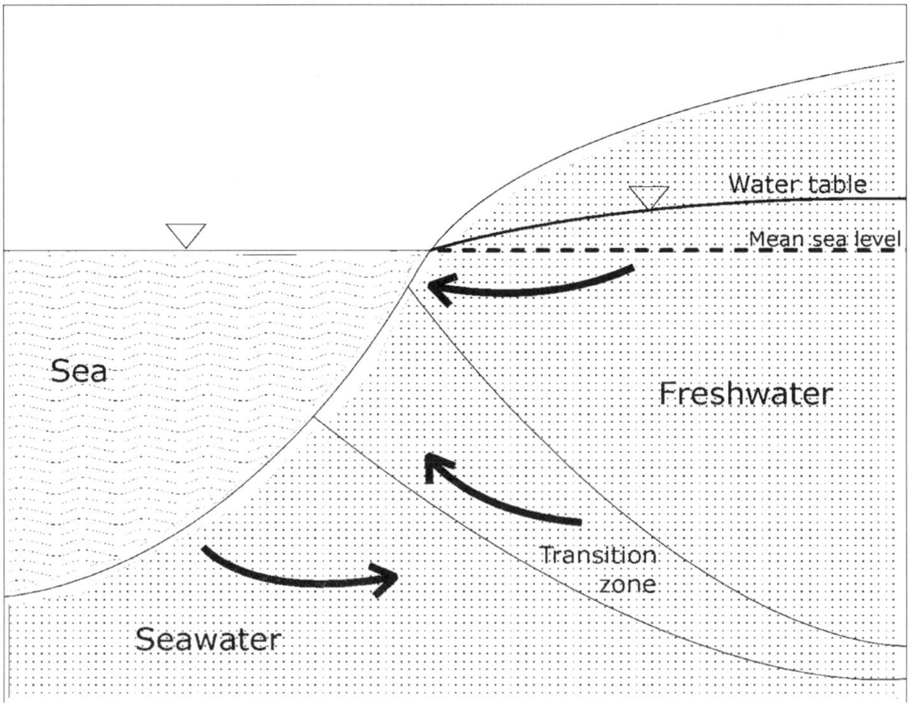

Figure 3.4. Schematic representation of seawater intrusion in a coastal aquifer (adapted from Sherif and Singh, 1996).

contaminant can cross the saturated zone until it reaches the bottom of the aquifer; if the density is lower, the contaminant will spread over the interface (water table) between the unsaturated and saturated zones. In the case of seawater intrusion, the contaminant is denser than freshwater, it normally enters an aquifer at some depth, and can make its way upward into unconfined aquifers and soil root zones aided by excessive groundwater pumping. For salts introduced overland from fertilizers, wastes, salty irrigation water, or salt ponds, complex fingering patterns can occur as the heavier saltwater percolates downward (Schincariol *et al.*, 1994; Simmons *et al.*, 1999).

3.4.1. *Seawater intrusion*

In a typical coastal aquifer, seawater and freshwater are separated by an interface across which a mixing or transition zone develops due to dispersive effects (Figure 3.4). Many modelling studies replace this mixing zone with a sharp front. If, in addition, one adopts the Dupuit assumption of predominantly horizontal flow, the problem of finding the interface may sometimes be solved in closed form. For the more general sharp interface problem, several two-dimensional numerical solutions have been published (e.g. Taigbenu *et al.*, 1984).

The Ghyben–Herzberg ratio can be used to determine the shape and position of the sharp interface under static equilibrium conditions (Sherif and Singh, 1996). For example, using 1.025 g/cm^3 as an average density of seawater, we find that the slope of the sharp interface is 40 times greater than that of the water table. If the water table drops 10 cm, the interface will rise a dramatic 4 m. The Ghyben–Herzberg ratio is also helpful in understanding the phenomena of saltwater upconing, which can occur, as seen earlier, in response to a pressure reduction from drawdown around a pumping well. If the pumping rate exceeds a critical level, the saltwater cone will reach the well. One of the weaknesses of the Ghyben–Herzberg ratio is that the saltwater interface intercepts the water table at the shore line, meaning that it does not allow for freshwater discharge to the sea.

When vertical flow and/or dispersion become important, or when the depth of the aquifer at the sea boundary is relatively large, the sharp front approach may not be adequate for describing seawater intrusion (Henry, 1964). As a result, several recent two- and three-dimensional models of seawater intrusion handle both vertical flow and dispersion (Frind, 1982; Huyakorn *et al.*, 1987; Voss and Souza, 1987; Galeati *et al.*, 1992).

3.4.2. *Soil and groundwater salinization*

Salinization of soils and groundwater has been observed in many cases where native vegetation is replaced with cultivated crops (Miller *et al.*, 1981). Practices such as fertilization and tillage can increase the salt content of soils and drainage water, but the most important cause is irrigation. Irrigation is an agricultural practice of major economic importance, especially in relatively warm and dry climates such as those which characterize the Mediterranean basin. Three processes can cause groundwater salinization in irrigated agriculture (Suarez, 1989):

- increasing concentrations of salt in the soil due to uptake of water by plants
- downward movement of salt in the unsaturated layer with leaching water
- intrusion of saline water as a consequence of pumping groundwater for irrigation.

Crops take up water and nutrients from the root zone, but salts responsible for salinization (NaCl and others) are usually left in the soil because they are not needed for plant growth. Since irrigation water very often contains some concentration of salt, this salt will tend to accumulate in the soil. When irrigation water is applied to cultivated fields, the process is seldom 100% efficient, in that a certain amount of water not consumed by evapotranspiration will percolate to the groundwater. As it flows though the layers of the cultivated soil, the water will increase its salinity due to dissolution of salts previously deposited along the soil profile.

Another possible impact of irrigation is the salinization of riverine and estuarine environments due to the diversion of water in irrigation canals (upstream saltwater migration). This diversion can reduce the discharge flow below a threshold (minimum vital flux), thereby promoting intrusion of seawater in coastal areas (Zalidis, 1998).

The process of salt accumulation in soils and groundwater can be considered as an example of agricultural diffuse pollution and can be modelled and managed using approaches similar to those used for other nonpoint source pollutants. In modelling the process, consideration of the unsaturated zone is important for analyzing salinization processes in detail and for investigating specific phenomena such as the water table response and salt accumulation at different rates of irrigation over long time periods.

3.5. Agriculture and groundwater in southern European countries and the Mediterranean

The Mediterranean is an area of transition between temperate environments and arid ones, where the relationships between human activities and natural resources can assume a wide range of dramatic changes. Typically, in temperate areas, water needs for various uses (agricultural, industrial, etc.) can be satisfied without excessive conflicts and thus environmental issues can play a relevant role in the debate about the use and conservation of water resources. On the contrary, in arid areas the debate tends to be more focussed on economic aspects and the environmental ones are usually neglected. Collin (1995) discusses these topics by defining the differentiation between the continuous water cycle found in the temperate environment and the discontinuous cycle of the arid areas. Very briefly, in the former case, rainfall can feed both surface and subsurface water compartments (with runoff and infiltration phenomena respectively) and subsequently groundwater can feed streams that have their origins in runoff along their courses. For such cases, correct exploitation of groundwater resources is usually posed in terms of avoiding quantitative and qualitative depletion: keeping the various compartments in equilibrium and maintaining the continuity of the cycle.

In arid areas, the water cycle is typically discontinuous; rainfall events are, in fact, episodic, not allowing significant infiltration to groundwater and responsible for only some of the runoff which could eventually be used to fill reservoirs. Groundwater resources which may exist are then disconnected from the other components of the water cycle. In such cases, groundwater may represent the only available water resource and its use can be viewed as the mining of a non-renewable resource; the main issue is then sustainable management and economic optimization of multiple uses.

It is possible to identify several common characteristics that describe the distinctive features of the Mediterranean basin. From a climatic viewpoint, rainy autumns and winters are followed by dryer springs and very dry summers; moreover rainfall distribution and patterns tend to favour high intensity events. These features tend to determine strong water deficits for the major cultivated crops, especially those with spring–summer cycles, but also make non-negligible runoff and leaching events during the wetter season possible.

The geomorphology of the Mediterranean is characterized by its relatively recent geological age. Besides relevant seismic and volcanic activities, this is responsible for there being few large plains in close contact with mountains and the sea, and thus, with the exceptions of the Nile, and also the Ebro, Rhone and Po rivers, the rivers are relatively short and the basins are small (Grenon and Batisse, 1989).

Three main types of aquifers have been identified by Attia (1998): carbonate karst, alluvial and multi-layered sedimentary. The first is the most prevalent, with very complex hydrogeology and with the largest systems being located on the northern side of the Mediterranean basin. The main alluvial aquifer systems are the Egypt valley and the Po valley in Italy. The latter type of aquifer is typically found on the southern side of the basin. Alluvial aquifers are, in general, characterized by high hydraulic conductivity and are in direct connection with the rivers that have a recharge function, and thus are more vulnerable to deep leaching than sedimentary aquifers, which are often protected by confining layers. Karst aquifers are more difficult to generalize due to their complexity and direct dependence on the specific fissure structures and layering.

From the above, it would seem that from the point of view of preserving groundwater resources in southern European countries and the Mediterranean, the problem of managing agricultural systems can thus be treated in the more general context of identifying ways of achieving sustainable agricultural development and of fighting the risk of desertification: overpumping, diffuse pollution and saltwater intrusion are some of the problems that should be dealt with. The Mediterranean, as a relatively dry area, is not among those (i.e. northern Europe) for which the generation of agricultural loads reaches the highest intensity, but the problem is still not negligible. In fact, the seasonal dynamics of climate and agricultural activities, combined with the vulnerability of the environment and with the scarcity of water resources, make the problem of agricultural diffuse pollution of greater relevance. Moreover, irrigated agriculture (implying intensive use of agro-chemicals) is still increasing in several areas, and thus also the relevance of the primary sector in the management of groundwater resources.

In summary, in the Mediterranean basin, groundwater resources are mainly impacted by agricultural activities as a consequence of:

- extraction of irrigation water all around the basin, but the southern part is of particular concern (mining of non-renewable aquifers);

- use or re-use of water of poor quality for irrigation, especially in those areas with lower rainfall inputs (salinization of soils and aquifers);
- use of water for irrigation with application systems of low efficiency, especially in those areas with relativcly good water supply from rivers or groundwater (pollution, but also salinization of aquifers);
- use of water for livestock rearing (salinization and nitrate pollution of aquifers);
- adoption of intensive cultivation systems (horticulture, intercropping, etc.) made possible by a favourable climate which makes irrigation water available (pesticide and nitrate pollution of aquifers).

From the viewpoint of the responses of the environment to the above mentioned impacts, the worst effects can be seen in cases of:

- consistent leaching phenomena induced by excess rainfall or irrigation;
- soils with high leaching potential (high content in sand, low cation exchange capacity);
- geologic conditions characterized by permeable (or fissured) layers;
- aquifers at shallow depth, with short and fast underground water paths;
- the presence of saline groundwater in proximity to freshwater aquifers.

Of course, the opposite conditions will favour the preservation of groundwater resources, but it has to be remembered that the complexity of the phenomena and of the interactions between the variables involved make it very difficult to generalize and judge real world intermediate situations.

Given the described framework of pressures, impacts and responses, various approaches can be taken towards setting up planning control programmes to cope with the above problems, but in general the following phases should be taken into account (Atkinson, 1987):

- problem definition;
- inventory and impact analysis;
- formulation of alternative control plans;
- comparative evaluation of control plans;
- selection and implementation of controls.

The first phase has been discussed above, and in what follows, the inventorization and analysis of the agricultural systems and of the mechanisms generating environmental impacts (phase 2) will be discussed, while in a later section, management of the existing farming systems and ways of putting strategies into place that could minimize environmental impacts (phases 3–5) are described.

3.6. Environmental impact analysis of agricultural systems

The inventorization and analysis of agricultural impacts on groundwater encompasses four tightly linked steps:

- identifying impact indicators;
- monitoring of the phenomena in the real world;
- modelling and simulating the observed phenomena with descriptive models;
- geographical analysis and simulation.

3.6.1. *Identification of impact indicators*

As stated previously, pollution phenomena are complex and consist of alterations in the state of many environmental variables. It is therefore necessary to choose those parameters and variables (from a long list of possible ones) which are most representative of various situations and to define their use as environmental indicators (OECD, 1994). An example of suitable indicators for the problems in question are concentrations of nitrate, chloride or pesticides in deep leaching from cultivated fields.

The chosen environmental indicators can be measured within monitoring activities or estimated by simulation models and used as quantitative information for the assessment of alternative hypotheses for agro-ecosystem management. For those management systems that are to be evaluated within a multidisciplinary approach, it is necessary to identify other indicators which can quantify possibly useful aspects: technical indicators (e.g. labour/machinery requirements) and economic indicators (e.g. cost of production factors).

3.6.2. *Monitoring of the phenomena in the real world*

When dealing with agricultural production processes, the monitoring phase consists of the acquisition of experimental data gained by means of agro-environmental trials in the field. Data are gathered that quantify the variations in pollution phenomena due to changes in cropping techniques. A typical monitoring activity consists of collecting samples of leaching water from neighbouring fields in which alternative cultivation techniques are used (e.g. different ways of fertilizing a given crop) (see for instance, Giardini and Giupponi, 1995). These activities can make available quantitative information about the magnitude of impacts and the possibility of proposing low impact alternatives. Parallel economic evaluations can propose cropping alternatives with a low environmental impact which are also acceptable in terms of farmers' incomes.

Monitoring activities should also be oriented towards the assessment of the qualitative/quantitative characteristics of the water resources to be used (e.g. networks of sampling wells). In this regard, it is important to point out that it is not feasible to monitor all aquifers and test all possible pollutants (many hundreds of agro-chemicals are used). Efficient sampling strategies for identifying critical site/pollutant combinations must therefore be set up (Holden, 1986). Moreover, it should be remembered that the orders of magnitude of the parameters in play are extremely variable. For example, nitrates in leaching waters can have concentrations between a few and hundreds of parts per million, while the total leachate of nitrogen can range between a few kilograms per year (dry climates, reduced fertilization) to values 10 or more times higher (humid climates, permeable soils, heavy use of livestock manure). The pesticide concentrations in the different environmental compartments are instead in the order of parts per billion and total annual losses are generally measured in grams per hectare.

Monitoring groundwater with the purpose of assessing the sustainability of current land and water management is also a complicated issue because of the difficulties in describing and interpreting the long-term dynamics of the system; the complexity can be even greater in the case of monitoring salinization because of the natural presence of salts in soils, geological layers and various water compartments.

The complexity of the phenomena and of the interactions between natural (e.g. rainfall patterns) and anthropic (e.g. fertilizer management) variables makes purely experimental approaches unsuitable. To be able to represent the functioning of the agro-ecosystems and to simulate their behaviour in different hypothetical situations, a modelling approach is therefore usually necessary (Van Keulen and Wolf, 1986).

3.6.3. *Modelling and simulation*

Environmental modelling integrating previous modelling approaches and experiences (mainly from the sectors of hydrology, chemistry and plant physiology) has become a sizeable branch of the agricultural sector. In the international literature, a number of models simulating the environmental impacts of agro-ecosystems, at various scales and for various purposes, have been proposed (Addiscott and Wagenet, 1985). Those models usually focus on the inefficiencies of agricultural plant production processes in particular (Giupponi, 1995), since the pollution phenomena are usually derived from the release of substances not utilized in the production process.

The availability of models calibrated for different environmental conditions is the basis for the simulation of alternative cropping system scenarios leading to proposals for sustainable land use (Giardini and Giupponi, 1994). Those scenarios may consist simply of various hypothetical cultivation systems, or more complex combinations of technical, economic and social alternatives, to be evaluated on the basis of their environmental implications.

3.6.4. *Geographical analysis and simulation*

Simulation models can be used to extrapolate experimental or field data acquired from short-term monitoring campaigns to much longer time periods (decades or even centuries) under varying environmental conditions (Ritchie, 1987). Analogously, once a model has been calibrated to some reference scenario or event, then with adequate territorial data, the model can generate spatial distributions (for instance, at a watershed scale) of any number of variables of interest. Using tools for digital image processing of satellite data (Richards, 1986) and for manipulating geographic information (Burrough, 1986), territorial analysis allows the creation of a systematic framework of information that provides a spatially and temporally distributed description of processes and phenomena, information that takes both natural and anthropic characteristics and the variability inherent in the area being studied into account.

It is of utmost importance that environmental studies make use of both simulation models and territorial analysis; indeed many methodologies exist for integrating the two. Simulations can tackle one or several phases of pollution (from generation to discharge to migration) but in general, it is the coupling of models and geographic analysis tools that produce three-dimensional descriptions of how the pollutant is transported, accumulated, and transformed, and of how the territorial system responds to various changes that may be introduced. Comparisons can thus be made between, for example, scenarios for land use and groundwater exploitation regimes. This will enable alternative territorial management policies to be identified. The complexity of the phenomena to be modelled and the intrinsic variability of the land and of land uses makes it extremely difficult to obtain accurate or absolute estimates of the magnitude of pollution under different combinations of soil, climate, crop management, and socio-economic conditions. Nonetheless, these methods can contribute, at least in their current form, to comparative assessments of alternative scenarios, and can provide substantial contributions to decision-making processes. When models are used for supporting planning activities and management decisions, great care should be given to the assessment and documentation of the various areas of uncertainty within the various phases of elaboration: data acquisition, algorithm development, modelling, etc. (Simonovic, 1997).

3.7. Modelling groundwater and agro-ecosystems

3.7.1. *Modelling approaches*

In the representation and simulation of subsurface flow and contaminant transport phenomena, and the agricultural processes that affect these phenomena, there are probably as many types of models, modelling philosophies, and ways of classifying models as there are sources and types of groundwater contamination (see, for example, Renard *et al.*, 1982; Mangold and Tsang, 1991). For instance, in the area of diffuse or non-point source pollution in agro-ecosystems, various models have been proposed that deal with different aspects of the plant–soil–climate continuum, some focused more on hydrological aspects, while others look at plant physiology or the chemistry of soil–water–plant interactions in more detail. For our purposes, we will consider three broad and not necessarily distinct classes of model or modelling approach:

- empirical and regression models, which express some heuristic or statistical connection between observed phenomena, or between inputs and outputs, without concern for the underlying physics, biology and chemistry, or the inner workings of the system;
- conceptual and analytical approaches, which greatly simplify the underlying physics (and other aspects) for reasons such as tractability, simplicity, and convenience;
- physically based or process-based models (sometimes also referred to as distributed models), which strive for as complete a description as possible of the underlying physics (and other aspects) within a deterministic or stochastic framework and within the limits of those processes and observations that are of interest.

To some degree, this hierarchy also encompasses the possible ways in which models can be used: the simplest models of the first category can serve as useful screening tools for identifying possible cause–effect links in a pollution incident; conceptual models can generate "what if" scenarios for impact assessment of, for instance, alternative cultivation practices or pesticide application rates; the third class is ideally suited to research and can be used to conduct exploratory simulations to test new hypotheses and parameterizations, for example. In practice, there is a great deal of blurring and overlap between the ways different models can be used, as will be shown later in an example of a regression/empirical approach and in a relatively detailed presentation of a standard process-based model for groundwater flow and transport.

We can include, as a fourth category to the three above, that of combined approaches, in the sense of composite models which use one approach for a given subset of processes or subsystems, and another approach for a second subsystem. Process models, based as they are on fundamental governing equations, are the most multi-purpose, flexible, and extendable of the approaches, though these comprehensive models are not without their limitations and drawbacks. Chief among these are over-parameterization and uncertainty, in the sense that most models have not been validated in all their detail, owing in part to a mismatch between model complexity and the level of data which is available to test and calibrate the models. In applying such models for large-scale studies, computational requirements can also be prohibitive.

3.7.2. *Implications of scale and variability for modelling*

The issue of variability and its links to problems of scale has been one of the dominant themes in hydrology for the past 15 years (Bloschl *et al.*, 1997). It is an issue that has a bearing on the enormous difficulties in both monitoring and modelling hydrological systems that have already been alluded to, and on efforts to quantify and minimize the uncertainty inherent in parameterizing, calibrating, and using process-based models in an operational and decision support sense.

On a practical level, considerations of scale and variability play a decisive role in the design of any field-based water resources study. For instance, addressing the interplay between groundwater pollution and European agricultural policy would appear to require a large-scale study site that embodies a wide range of processes and generates inflows and outflows of social and economic significance (affecting a large population, or representing a sizeable fraction of total agricultural production, for instance). But a large-scale study requires enormous amounts of data, and may not be amenable to a process-based modelling approach. Studies restricted to smaller regions can nevertheless be useful in evaluating the relative importance of pollutant transport and transformation mechanisms and in assessing the effectiveness of various remediation strategies. The results of such studies can then be implemented at larger scales, with consequent implications for national or European agricultural practices. We mention a few examples that illustrate this:

- The important buffering effect of riparian forests or wetlands on nitrate discharges can be readily studied using a process-based model applied to a hillslope or small watershed.
- Models of agro-ecosystems deal with the plant–soil–climate system at various scales (both spatial and temporal), but in general, the limits of the system (at which estimates of pollution loads are made) are defined as the bottom of the root zone, for leaching, and as the edge of the field for surface runoff (see Figure 3.1).
- The salinization cycle, which involves seawater intrusion or saltwater upconing in deep confined aquifers, extraction of this saltwater by pumping, use of the pumped water for irrigation, and consequent accumulation of salt in soils as crops take up mostly freshwater, can be modelled in a two-stage process firstly using a three-dimensional aquifer model and then a one-dimensional vertical flow and transport model with a simple treatment of water and solute (salt) uptake by plants.
- Physically based numerical models can be used to individually simulate non-point source pollution over many small basins or fields that are representative of a larger region of interest (country or continent, for instance). The simulation data can then be statistically analysed to obtain the spatial and temporal patterns of solute concentrations, and the results aggregated to enable quantification of large-scale pollution characteristics.

3.7.3. *Empirical and regression models: an example for regional water quality assessment*

We describe an example from the recent literature that deals with large-scale watershed pollution, predominantly non-point source nutrients. The example will be used to describe some of the fundamental hydrologic processes involved, the data requirements, and the important inputs and outputs, as well as to elucidate this interesting regression modelling approach. Smith *et al.* (1997) developed a regression methodology to derive total phosphorus (TP) and total nitrogen (TN) concentration and transport rates for the entire conterminous United States. The model makes use of a digitized network of stream reaches and associated land surface polygons, and data from 414 water quality stations located near the outlets of selected watersheds that are part of the U.S. Geological Survey's National Stream Quality Accounting Network (NASQAN). The statistical regression model 'relates measured transport rates in streams to spatially referenced descriptors of pollution sources and land surface and stream channel characteristics'.[1]

For the total nitrogen analysis, the authors considered five β factors (sources) – point sources, fertilizer application, livestock waste production, atmospheric deposition, and non-agricultural land; eight α factors (land surface characteristics) – temperature, slope, soil permeability, stream density, wetland, irrigated land, precipitation, and irrigated water use; and three δ factors (stream channel characteristics) – slow (stream flow $Q < 1000\ \text{ft}^3/\text{s}$ or $28.3\ \text{m}^3/\text{s}$), medium, and fast ($Q > 10\,000\ \text{ft}^3/\text{s}$) flow classes.

The results obtained upon fitting the model by non-linear least squares estimation suggest that for total nitrogen, the most significant land surface parameters are temperature, soil permeability, and stream density, and that the channel decay coefficients decrease as the stream size (or flow rate) increases. High temperatures increase the rate of denitrification, and therefore can be expected to decrease the delivery of TN to streams. Highly permeable soils are expected to allow more contaminants to enter the subsurface, where the contaminants are then subjected to additional degradation processes and longer travel times to the stream channels. Stream density (or drainage density) is the ratio of channel length to watershed drainage area, and thus higher stream densities imply shorter overland travel distances for contaminants before the stream is reached.

The model developed by Smith *et al.* (1997) is empirical, but nonetheless has a number of very appealing features:

- Its simple structure and statistical nature allow investigation of a wide range of factors that affect (or may affect) TN and TP transport rates.
- It is parameterized according to quantities that can be easily measured (although this also has drawbacks in that these parameters can lump together a number of physical effects, thereby making it difficult to identify or isolate important underlying processes).
- An enormous amount of data can be analysed (in fact a large amount of data is normally a requirement for obtaining statistically significant results).
- The model is not physically based, but because it relates water quality to spatially referenced watershed attributes, it has some interpretative capabilities that provide a link between the descriptive and explanatory aspects of assessment. This can be used directly in water resource decision-making

frameworks, or it can yield insights to guide the construction of more complex process-based models.

The authors conclude that the developed methodology and model is "an important adjunct to data collection in regional water quality assessment programmes as a means of addressing the problems imposed by limited sampling resources, network bias [that arises from the fact that specific or known pollution sources and regions are often monitored more thoroughly than others], and basin heterogeneity".

3.7.4. *Process-based models*

The standard mathematical model governing water flow and chemical transport processes in porous media is based on the partial differential equations of fluid mass and momentum balance and of solute mass balance. These equations are described here in their most general three-dimensional form; depending on the application, simplification to one or two dimensions is common practice. These equations can be extended and elaborated in a variety of ways to enable treatment of numerous specialized or more complex cases (Addiscott and Wagenet, 1985; Parker, 1989; Ségol, 1993; Gallo *et al.*, 1996; van Genuchten and Simunek, 1996; Bixio *et al.*, 1999). In the following section, an extension of the basic model to the problem of saltwater intrusion will be demonstrated; other extensions include formulations for treating nonequilibrium chemical transport, biodegradation, multi-species and multiphase phenomena, radionuclide decay chains, preferential flow (fractures, macropores, cracks), water uptake from plant roots, coupled heat transfer and moisture flow, and coupled subsurface and overland flow (surface runoff and channel routing).

Flow in variably saturated porous media is governed by Richards' equation.[2] This equation is strongly non-linear due to pressure head dependencies in the relative hydraulic conductivity and general storage terms and must be solved iteratively using linearization techniques such as that of Newton-Raphson (Paniconi and Putti, 1994). These dependencies have been extensively studied and are expressed through semi-empirical constitutive or characteristic ratios describing the soil's hydraulic properties (see, for example, van Genuchten and Nielsen, 1985). The transport equation[3] describes diffusion, dispersion and advection processes, as well as simple chemical reactions (linear equilibrium adsorption and radioactive or biodegradation decay). For the numerical discretization of the flow and transport equations, a standard finite element Galerkin scheme is used, with tetrahedral elements and linear basis functions, complemented by weighted finite differences for the discretization of the time derivatives. For an introduction to finite element techniques in engineering and groundwater applications see Zienkiewicz (1986) and Huyakorn and Pinder (1983). The finite element discretization yields large sparse systems of equations which are solved using efficient preconditioned conjugate gradient-like methods (Gambolati *et al.*, 1996).

To complete the mathematical formulation of the flow and transport problem, initial and boundary conditions need to be specified. Initial conditions consist of the state of the system at the start of the time period being simulated. When the model is discretized, this information is given in terms of pressure head and concentration values at each node in the interior of the domain where the domain corresponds to the system (aquifer, watershed, etc.) being modelled. Boundary conditions are instead required for the entire simulation period, but only at the nodes which constitute the boundary of the simulation domain. This information can take the form of assigned pressure head and concentration values (Dirichlet type), prescribed flux values (Neumann type), or a mixture of these (Cauchy type). Different segments of the boundary can have different types of boundary conditions.

In addition to the various parameter values that need to be assigned a priori to the model (saturated hydraulic conductivity, porosity, dispersivities, coefficients in the storage and relative conductivity expressions, etc.), the initial and boundary conditions constitute another set of model inputs. Detailed or accurate values for these inputs are not always readily available or easily measured, and approximations such as assuming the porous medium to be homogeneous, are often necessary. In the case of initial conditions for the transport model, selecting the start of the simulation period to be prior to the occurrence of contamination means that zero concentration can be initially assigned to all nodes. For boundary conditions, knowledge of the geological and hydrographic features of the simulation domain is important, as is careful delineation of this domain. Rivers, watershed divides, a layer of bedrock underlying an aquifer, and other such natural boundaries can all be treated relatively easily.

3.7.5. *Coupled flow and transport model of saltwater intrusion*

The mathematical formulation for the three-dimensional finite element model that treats density-dependent variably saturated flow and miscible (dispersive) salt transport is developed as an extension of the basic equations described in the previous section. The formulation and procedures form the basis of the CODESA-3D (COupled variable DEnsity and SAturation) model (Gambolati *et al.*, 1999).

The mathematical model of density-dependent flow and transport is expressed in terms of an equivalent freshwater head h, defined as $h = \psi + z$ where $\psi = p/(\rho_o g)$ is now the equivalent freshwater pressure head, p is the pressure, ρ_o is the freshwater density, and g is the gravitational constant (Huyakorn *et al.*, 1987; Frind, 1982; Gambolati *et al.*, 1993). The density ρ of the saltwater solution is written in terms of the reference density ρ_o and the normalized salt concentration c: $\rho = \rho_0(1 + \varepsilon c)$ where $\varepsilon = (\rho_s - \rho_o)/\rho_o$ is the density ratio, typically $\ll 1$, and ρ_s is the solution density at the maximum normalized concentration $c = 1$. Depending on the application, ρ_s can represent, for instance, the density of seawater or of the solution of a salt pond where the concentration is highest. The dynamic viscosity μ of the saltwater mixture is similarly expressed as a function of c and a reference viscosity. With these definitions and constitutive relationships we can derive the model describing saltwater intrusion phenomena.[4]

Coupling in this model is due to the concentration terms that now appear in the flow equation and the head terms that appear in the transport equation via the Darcy velocities. In the simpler case of non-density-dependent flow and transport, the system is coupled only through the head terms in the transport equation. In this case there is physical coupling, but mathematically the system can be reduced (decoupled) and solved sequentially, first the flow and then the transport equation, without iteration. For the density-dependent case, the system is irreducible and any sequential solution procedure requires iteration. The importance of coupling, and the degree of non-linearity now present in the transport equation (in contrast to the basic model, where only the flow equation was non-linear for the case of unsaturated porous media), are expected to decrease as the density ratio ε decreases or as dispersion becomes dominant (Putti and Paniconi, 1995).

3.7.6. *Parameter estimation and model calibration*

Obtaining sufficient and reliable input data to assign to the various parameters and initial and boundary conditions of simulation models can be an onerous task. Given that inevitably in models of natural systems such as aquifers and watersheds, the available data tend to be scarce or inaccurate, there has been much research devoted to methodologies for parameter estimation and model calibration. The procedures commonly used range from ad hoc trial and error methods to sophisticated mathematical/statistical inverse algorithms. In a recent and exhaustive study comparing seven inverse approaches to estimate parameters for flow and solute transport models (Zimmerman *et al.*, 1998), the authors concluded that "the most important factor for achieving a successful solution was the time and experience devoted by the user of the method". This underlines an important point that applies not only to inverse methods, but more importantly to groundwater models themselves: comprehensive physically based numerical models are powerful tools, and are increasingly being equipped with user-friendly graphical interfaces and instant post-processing utilities (plots, maps, summary reports, etc.), but their proper use for analysis and decision support will continue to require on the part of the user some knowledge of the underlying processes and their interactions, of the mathematical representation of these processes via equations and parameters, and of the particular features of the computer implementation – these are elements that can make the difference between a successful simulation and a meaningless one.

3.8. The role of geographical information systems (GIS)

3.8.1. *Model inputs, outputs, and derived information*

The instruments that are used for monitoring and measuring physical, chemical, and biological processes in a natural system such as an aquifer or watershed provide input data for simulation models. The models recast this input data in the form of boundary conditions and other external forcing terms (such as pumping wells), initial conditions, and values for various physical and numerical parameters. The models then produce outputs such as multi-dimensional fields of basic and derived variables at selected times (pressure heads, pollutant concentrations, water table levels, moisture contents, etc.), hydrograph output showing the flow rates at a designated point (a catchment outlet or a monitoring station), and diagnostic output summarizing the performance of the model run. This input and output information reflects the status and evolution of atmospheric, soil, and subsurface resources, and is normally utilized by scientists in raw form. Roughly speaking, a single number gives the value of a specific parameter or state variable at a given point in time and space. This is not normally the form or nature of information required by a decision maker, resource manager, or policy analyst, for a number of reasons:

- The parameter or variable may not be meaningful or directly utilizable by the policy analyst – raw data need to be converted to another form.
- The data may not be at a space or time scale consistent with the needs of analysis – raw data need to be (dis-)aggregated.
- The analyst requires information that is represented or contained in a number of parameters/variables – policy information needs to be derived from raw data.
- The information required is not just process-based or scientific data, and must be combined with other data (demographic, economic, medical, etc.) – raw data need to be integrated with other data sources and types.

An important issue that arises, therefore, is that of transforming model inputs and outputs into format and content useful for non-scientist end-users, in the form of maps, indicators, indices, forecasts, scenarios, statistics, and so on. Given that much of the raw data is spatio-temporal in nature (think of water quality measurements from a monitoring network, a satellite image of land cover, or a calculated map of salinity isolines), it is natural to use a GIS as a primary tool for data organization and transformation, and indeed the role of GIS in hydrology and water resource studies is well recognized (Leipnik *et al.*, 1993; Maidment, 1993; Moore *et al.*, 1993).

3.8.2. *What is GIS?*

The trend of growing awareness and concern for the environment and the need to make rational choices to deal with environmental issues has gradually brought geographic information, generally intended as "information which can be related to a location on the Earth, particularly information on natural phenomena, cultural, and human resources" (AGI, 1991), to be considered as the basis for an improved understanding of many of the problems afflicting our planet. In parallel, the complexity and the heterogeneity of geographic information, as represented by maps, remote sensing imagery, monitoring network logs, socio-economic data and so on has required the development of adequate technologies for its representation, processing, and management (see UNEP, 1999 for an overview). Since the 1960s GISs have evolved in response to these needs and in tandem with the evolution of related technologies such as database management systems, visualization, geostatistics and remote sensing (e.g. Ehlers *et al.*, 1989). As an evolution of cartography-based applications, GISs have naturally assumed a role of primary relevance in domains such as natural resource management and land planning. At the same time, GI technologies have also developed within domains such as facility management, tourism, telecommunications and transport.

A general definition of a GIS which highlights its key characteristics can provide a useful starting point for describing the possible role of such a technology in policy support for environmental or hydrological applications. In the various definitions of GIS given in the literature,[5] common features that emerge are those of the interaction between institutions, either represented by analysts or policy makers, the importance of data, and the availability of an array of support tools. Only an appropriate balance between these components can insure the further evolution of GIS from a strictly technical solution to a tool accepted in the policy process.

Environmental data is generally complex, voluminous and characterized by heterogeneities and discrepancies due to often ad hoc data acquisition over time. Analysis tools and methodologies which are consolidated within a single discipline may need to be adapted when brought into a multi-disciplinary arena. In the past few years these issues have been clearly identified and recognized as a priority by the major subjects responsible for the acquisition and management of geographic information: national mapping agencies, data providers (e.g. the European Space Agency), and communities of software developers and users such as the OpenGIS consortium (OGC Technical Committee, 1998), for example. Concerning data, the adoption of accepted standards for their content and description (or metadata) should enable providers to produce data sets with known quality and harmonized across boundaries, spatial scales, and geographic projection, allowing a seamless merging of the most diverse sources of geographic information.

The availability of metadata is the basis for the simplification and enhancement of directory and data retrieval systems which are fundamental for the exchange of information between the numerous players involved in the analysis and definition of policies related to environmental problems. The scenario which is envisioned in the evolution of GIS is strongly influenced by the development of increasingly powerful information and communication technologies (European Commission, 1998), and is represented by the creation of spatial data infrastructures, both at a national and trans-national level, which should allow geographic data and knowledge to be located and shared.

3.8.3. *GIS and environmental/hydrological modelling*

Understanding the spatio-temporal behaviour of hydrological processes and state variables at large scales involves the use of many different types of data, obtained from field measurement, remote sensing, digital terrain models, and numerical simulation. A GIS support is particularly valuable at the modeller's level. For example, the design of a numerical grid representing an aquifer for application of a simulation model can be automated and more directly linked to the mappable features in the study area. This makes the process more intuitive and relieves the user from tedious and error-prone processing tasks (Kuniansky and Lowther, 1993), while improving the accuracy of the description of the site under examination. At the same time, the fact of undertaking a modelling study in a GIS context provides a basis for the simplification of the interaction between the different players involved (data providers, modellers, and decision makers) through the establishment of a common data structure.

With the complexity of models and the variety and volume of data that needs to be processed, pre- and post-processing tasks related to modelling efforts rely not just on GIS, however, but on a host of other software tools such as scientific visualization systems, image processing software, and database management systems. Combining these data, models, and tools into a robust and user-friendly system is a research topic that has seen approaches ranging from so-called loose integration to tight integration (Batty and Xie, 1994; Livingstone and Raper, 1994; Nyerges, 1994; Paniconi *et al.*, 1999). Another approach to GIS-model integration is to move from modelling linked to GIS to modelling within GIS. This is generally achieved by implementing fundamental modelling primitives as intrinsic GIS functions, such as the advection–dispersion equation for groundwater transport, or by characterizing spatial response functions via time-area diagrams, for example (Maidment, 1996). The outcome of such an approach is clearly dependent on the type of model being considered, and several models have already been successfully integrated within different GIS packages.

Three examples of GIS-integrated land surface or subsurface models for hydrological and agricultural applications are the DRASTIC methodology for groundwater vulnerability mapping (Merchant, 1994) and the AGNPS and ANSWERS models for non-point source pollution modelling (Wilson, 1996). A fourth example is BASINS (EPA, 1998), an integrated watershed-based modelling system for water quality assessment and analysis of point and non-point sources of pollution. BASINS is a tool aimed specifically at agencies responsible for pollution control and water policy and regulation. In particular, it is intended as a support tool for the establishment of total maximum daily loads (TMDLs) for a wide variety of pollutants, as required of all US states over the next 15 years by the Clean Water Act. One of the distinctive features of BASINS is that it integrates not only software tools (including GIS and simulation models), but also data. Indeed, data is a key component of BASINS, with 1–2 gigabytes of geographic and environmental data per EPA Region distributed as part of the software package (there are 10 EPA Regions in all). This approach ensures compatibility and consistency between BASINS implementations within different watersheds, and encourages adherence (for any additional data introduced in a local application) to established data and metadata standards such as that of the FGDC (Federal Geographic Data Committee). The data supplied with the BASINS system is derived from a wide range of US agencies (NASA, USGS, EPA, etc.) and includes cartographic, land use, soil, stream, digital elevation, meteorological, water quality and pollutant loading data.

While consistent progress has been achieved during the past years in the development of integrated GIS/modelling solutions, it has been observed that if we accept the concept of GIS as a methodology for handling spatial location and interrelationships, while environmental modelling handles system states and dynamics, full integration between the two will always suffer from a number of representational compromises. In this regard, substantially different strategies that exploit object-oriented modelling of geographical features have been proposed (Raper and Livingstone, 1996; Crosbie, 1996).

Whichever the technical solution adopted to integrate geographic databases, GIS functionality, and ancillary tools with simulation models, the increased usability of the resulting system must be adequately supported in order to avoid improper or inexpert use (which paradoxically may entail an increasing risk as such tools become more sophisticated yet easier to use). This has led to the suggestion of yet another strategy for GIS model integration, plausible in situations where the use of sophisticated analytical tools is not warranted by the amount or quality of data available, whereby reliance on quantitative estimates is replaced by a qualitative understanding of the pattern of hydrological response and simple GIS-based reasoning is used to assist in the decision-making process (Grayson *et al.*, 1993).

3.8.4. *GIS and policy support*

As already mentioned, the data from environmental monitoring networks and the results from simulation models must be presented in a form that is understandable and effective for the policy maker. In this context, GIS can play an important role in, for example, conflating data from the local level (e.g. provincial and regional) to the global level, re-aggregating cell-based information to indicators referred to administrative boundaries, deriving complex maps by means of spatial analysis (overlay, buffering, map algebra), generating tabular and chart reports (a natural complement to the map-based information normally provided by a GIS), and supporting what-if scenario simulations (possibly designed in a way that avoids the dysfunctional separation of the roles of analyst and decision-maker (Nyerges, 1994)).

Notwithstanding the highly intuitive nature of map-based representation of information, geographic information systems should not, however, be considered as the sole or dominant tools to be proposed in a decision support system for a policy maker. Rather, GIS should be placed at the same level as other information management tools, together with visualization (in general) and databases (Peirce, 1998). Identifying the priority issues raised by the need to transmit information from scientists and analysts to policy makers leads in turn to the definition of strictly technical problems, where GIS play an important role. However, since any tool applied in the definition of a policy acquires a political connotation in itself, we should also consider whether the application of GIS to the policy process differs from the application of other more traditional tools or models, such as those used for economic planning or welfare policy analysis. With respect to this, King and Kraemer (1993) suggest that geographical information is more likely (compared with other modelling bases) to be accepted by different parties as a boundary object, while at the same time, the very breadth of GIS applicability to policy problems makes it likely that GIS will be drawn into many different kinds of policy debates.

An example of the change in perspective from modeller to analyst or policy maker can be of use in demonstrating how and, depending on the specific issue considered, to what extent GIS support can be of use in transmitting information between these players. Let us consider a sample set of questions which the policy maker might pose, in relation to a site that is at risk of saltwater intrusion, in order to make rational judgements regarding possible remediation, conservation, or regulatory actions:

1. Which are the zones or pumping wells exceeding or at future risk of exceeding maximum allowable salinity levels for a variety of water uses (urban, industrial, agricultural)?
2. Is the freshwater–saltwater interface advancing, receding, or in equilibrium?
3. What are the threshold or optimal irrigation rates for avoiding salt buildup in soils and to enhance flushing?
4. Can a localized aquifer recharge strategy that remediates extreme salinization be identified?
5. Can regional (aggregate) pumping location and rate regulations be designed that balance aquifer use and conservation pressures?
6. What are the likely sources or origins of the salinity in the soils and aquifers?
7. Is the monitoring network adequate? How can it be improved? (distribution of monitoring stations, frequency of data acquisition, what data has to be collected?). At what cost?

Table 3.1 lists the input and output data needed and generated by a generic

Table 3.1. Model input, output and calibration data for cases of seawater intrusion

Hydrometeorological data (input)	(a) topographic and geomorphologic data (channel networks, subcatchment units, etc.)
	(b) soil characteristics and hydraulic functions
	(c) rainfall/evaporation rates
	(d) overland runoff data
Soil and land use data (input)	(e) land use and agricultural data (e.g. crop salt tolerances, root uptake rates)
	(f) irrigation data (application rates and salinity levels)
Hydrogeological data (input)	(g) geological data (statigraphy, characteristics of the aquifers, aquitards, lenses, fractures, faults)
	(h) saline deposits and formations
	(i) porosity
	(j) hydraulic conductivity
	(k) aquifer storativity
	(l) dispersion (dispersivity coefficients)
	(m) density ratio
Other data (input)	(n) any other information concerning hydro/geo/pedologic heterogeneities
Aquifer use (input)	(o) location of pumping wells
	(p) pumping rates
	(q) aquifer recharge data
Monitoring data (model input, calibration, updating, and verification)	(r) streamflow rates and quality
	(s) piezometric data including salinity levels
	(t) water table levels
Model output	(u) salt concentrations
	(v) saltwater/freshwater interface (position, sharpness, dynamics)
	(w) soil moisture content
	(x) groundwater pressure heads
	(y) groundwater pressure gradients
	(z) groundwater velocities

process-based seawater intrusion groundwater model. For each of these questions or issues, Table 3.2 indicates the data that is pertinent to the issue (in this table other refers to non-process or scientific data, such as socio-economic data). Given that virtually all the types of data presented in Table 3.1 are characterized by spatial variability (even though a number of the parameters listed are, in practice, assumed to be constant, but this has mainly to do with simplifications or deficiencies in the model structure or in the data acquisition), there is a clear need for GIS technology to adequately process the information. Moreover, the input data are seldom obtained from a single source, thus data integration is another critical task that can be adequately addressed by GIS. Finally, the many-to-one mapping between input/output data and a given policy issue that is apparent in Table 3.2 gives a good idea of the extent to which model data need to be aggregated and transformed, via GIS and other tools, in order to address relevant policy issues.

Table 3.2. Model input, output and calibration data (see Table 3.1) relevant for addressing the seven policy issues or actions described in the text. 'Other' refers to non-process or scientific data (e.g. socio-economic data)

Policy issue	Input																Calibration			Output						Other
	a	b	c	d	e	f	g	h	i	j	k	l	m	n	o	p	q	r	s	t	u	v	w	x	y	z	
1					•									•					•		•						•
2			•							•		•				•						•			•	•	
3	•	•		•	•	•								•							•		•	•			•
4							•			•	•						•		•	•		•		•	•		•
5	•		•				•	•	•	•	•	•	•	•	•	•								•	•	•	•
6						•	•	•					•		•	•	•	•	•		•	•			•	•	
7						•								•	•	•		•	•	•	•			•			•

3.9. Identifying sustainable agricultural systems

Recognizing the existence of environmental impacts of agricultural origin (salinization or pollution) normally implies a public will to intervene to minimize or eliminate the observed negative phenomena. The possible strategies of intervention obviously differ depending on the type of impact and the socio-economic and environmental contexts, but, in general, policies to combat both salinization and pollution of groundwater resources can be threefold (Umali, 1993):

- issue measures to make efficient use of water;
- sustain the adoption of environmentally sound production methods;
- encourage wider use of environmental impact assessment of water exploitation plans.

This section aims to develop a more profound understanding of the interface between technical aspects and that of decision making and policy design and development.

In general, it is more efficient to prevent environmental impacts, or reduce them to an acceptable level, at their sources (i.e. the generation phase). For such cases, the most widely used approach is that of internalizing the costs of control measures (FAO, 1996) by creating an economic incentive for farmers to adopt more efficient production processes, and in particular, so-called eco-compatible cropping systems (i.e. with a low environmental impact). This can be done through agricultural policies that create conditions of economic advantage for these systems, or with appropriate land planning measures or regulations (Giupponi and Rosato, 1995).

It is usually more difficult to intervene in the transport phase, especially when dealing with subsurface fluxes of pollutants (i.e. impacts on groundwater). Nevertheless, pollution control actions on surface waters, which may have indirect effects on groundwater return flux, can be carried out: ways of doing this include intervening in the riparian vegetation (e.g. plantation of buffer strips), or oxygenating the water body by means of waterfalls, or constructing wetlands for phyto-remediation (Welsh, 1991). When pollutants reach the final receptor (aquifer, lake, lagoon, etc.) fewer options are available: these may include treatment plants for cleaning the water and restoring it to a quality adequate for specific needs (Kinzelbach and Schafer, 1993); making water suitable for drinking, for example.

In the context of making technical knowledge available for defining policies in the agri-environmental sector, the information gained on the phenomena generating environmental impacts is a precondition for setting up strategies and interventions to control the diffuse loads, with the aim of minimizing their magnitude (see, for instance, Yurdisef and Jamieson, 1997). A possible methodological approach for such purposes, recently developed and applied to the area of the Venice Lagoon watershed (Giupponi and Rosato, 1998, 1999), consists of the following main phases:

- definition of evaluation indices;
- definition of alternative scenarios;
- calculation of indices;
- multi-disciplinary evaluation and decision-support.

3.9.1 *Definition of evaluation indices*

Assuming that an adequate set of indicators has been identified during the phase of inventory and analysis, adequate methods and data for describing the agricultural management systems should be already available. The aims of the intervention having been precisely defined within the broad context of pollution control, it is then necessary to define suitable algorithms to apply to the values of indicators in order to obtain concise indices on which to base the evaluation of the alternatives (Giupponi, 1998). The alternatives are, in general, made up of possible scenarios for agricultural land use derived, for example, from the adoption of different government and planning acts or from the adoption of possible new technologies. An environmental indicator (a time series of values of pollutant concentrations, for example) can thus lead to the calculation of various kinds of concise evaluation indices to quantify (for instance) the risk for aquatic life in surface water or the suitability of aquifers for supplying drinking water. According to the aims and priorities of the interventions, the values of indices can be elaborated in various ways to be used in multi-disciplinary evaluation (e.g. cost–benefit analysis or multi-criteria analysis) together with the results of the economic and technical–agronomical approaches.

3.9.2. *Definition of alternative scenarios*

Variations in socio-economic and policy scenarios usually determine changes in farmers' decisions in respect of production processes and management techniques, decisions which are reflected in changes in land use at the territorial scale, and ultimately, in changes to potential impacts. New policies should therefore be based upon in depth knowledge of agricultural systems and their behaviour in response to external driving forces (i.e. policy and market contexts), forces to which the agricultural sector has shown itself to be quickly responsive, and which the history of the Common Agricultural Policy can provide many examples of.

In the past, the introduction of incentives per yield unit for some crops has pushed farmers to increase both the surface under subsidized crop cultivation and to intensify cropping techniques (e.g. higher fertilizer rates). More recently, incentives per unit of cultivated surface have increased the areas planted with those crops, but not the cropping inputs. These two policies for supporting farmers' incomes are evident examples of socio-economic actions that can cause dramatic and unexpected environmental effects. Changes in the ratios of costs and incomes among alternative crops are determining the choices made by farmers, who are varying cultivation techniques, crop allocations within the farm, or even completely abandoning or taking up new crops and cultivars as a consequence.

Many factors affect farmers' choices: farm size and typology, labour availability, risk aversion, and so on. Adequate surveys of farmers' strategies and behaviour in a given area must therefore be carried out to build farmer decision models and then to forecast the possible variations in land use in the different parts of the land. On this basis, alternative land use scenarios could be identified and implemented: this would enable the building of a framework within which models to feed the evaluation procedure with quantitative data (agri-environmental indices of agricultural systems) could be developed.

3.9.3. *Calculation of the evaluation indices*

Once the parameters that describe the types of land use are known (proportion of hectarage of the different crops, cultivation techniques, etc.), it is possible to calculate the associated impact indices and compare the environmental effects of the proposed alternatives (Giupponi and Ghetti, 1996). By operating within the context of a geographical information system, it is also possible to manage huge data sets (many crops in combination with different agronomic techniques, different environments, etc.) in a spatially distributed way. This allows not only a representation of the estimated magnitude of pollution phenomena, but also spatially explicit systematic and quantitative comparisons between the alternative scenarios to be very efficiently obtained: for example, maps of the differences in impact associated with two or more alternative policy scenarios (Giupponi and Rosato, 1999).

3.9.4. *Multi-disciplinary evaluation and decision–support*

When there are adequate information bases for production processes and environmental phenomena on the one hand, and for territorial features on the other, it is possible to assemble the results of environmental modelling and other approaches and to carry out multi-disciplinary evaluations. As agronomic, economic and environmental approaches generally show contracting trends in the values of the indices, multicriteria analysis approaches towards formulating the choices and informing the decisions are often adopted. In some cases, proper decision–support systems (DSS) can be produced as dedicated softwares. These systems, which are based on quantitative information on the foreseeable effects of choices, can be utilized by the authorities or individuals in charge of management at various levels to support their own choices (UNITAR, 1995). It is possible, for example, to compare the environmental effects of different cropping systems in the various parts of the territory in order to highlight possible conflict situations, and to plan interventions to ensure that higher impact cropping systems are not practised in the more vulnerable areas.

To be efficient, a DSS must have an extremely simple interface with the user and must present the results of scenario simulations within a very short time. Only in this way can the user (public administrator, extension officer, professional) effectively gain an advantage from the knowledge of the experts in the various disciplines that has been implemented within the system and encourage him to compare his own opinions with the indications that come from the DSS. The results of the elaboration must also be as well documented as possible to be able to refer to the criteria adopted, the decisional mechanism used and the degree of uncertainty involved.

3.10. Final remarks

There are many ways of classifying groundwater and soil contaminants and the nature of their sources (Domenico and Schwartz, 1990): according to localization (point source for fixed, small scale sources that produce well-defined plumes (landfills or a leaking pipeline, for example) and non-point source for diffuse, larger scale contamination (herbicides applied to farmland areas or runoff from urban centers, for example)); according to origin (industrial, military, agricultural, urban, and natural); according to chemical or biological properties (radioactive substances, trace metals, other inorganic species, nutrients, organic chemicals, and microbial pollutants). We have focused on a small but important set of nonpoint source contaminants that are commonly (though not exclusively) associated with agricultural practices encompassing pesticide and fertilizer use, groundwater pumping, irrigation, and animal farming and grazing. These contaminants are nutrients, pesticides, and saltwater. The fundamental processes that determine the production, fate and impact of a groundwater contaminant have been described, and we have given an overview of the basic mathematical equations governing fluid flow and solute transport in porous media, developing in more detail the numerical model for saltwater intrusion. This process-based modelling approach has been placed in the context of other approaches, with examples illustrating how various models can be used in practice and the precautionary steps needed to avoid misusing them.

Some of the important issues and concepts that accompany efforts to gain a better understanding of hydrologic behaviour and an improved capacity to model and predict this behaviour have been addressed, with particular emphasis given to the problem of transforming model inputs and outputs to information that is useful and useable for policy analysis and resource management. To this end, it should be emphasized that the theoretical knowledge base for understanding and describing agricultural impacts on groundwater is well developed. Technologies such as simulation models and GIS can, in this regard, provide a bridge to real world situations and problems, and allow the efficient organization of information management systems and, beyond this, the construction of effective decision support systems. An extensive list of possible technical solutions is available, as we have shown, from which, using these information technologies, rational choices can be made as to those most suitable for any given situation.

In the implementation and application of modelling, GIS, and DSS frameworks as resource policy and management tools, it is important to identify the appropriate approach (more than the specific model or software package) for the problem at hand (type of pollution, scale of application and impact, data requirements and availability, end-user needs such as prediction or hypothesis testing, and so on). In addition to ongoing progress in the development of technological systems, this will require closer interaction between the disciplines. Specific research areas that can benefit from such interaction include:

- greater ease of integration and exchange of data deriving from monitoring and simulation and needed for defining and evaluating strategies and policies;
- further evolution in the development of software tools in terms of user interfaces and the processing and presentation of observation data and simulation results;
- study of methodologies for closer coupling between physical and socio-economic models, beyond current techniques based on introducing simple constraints in an optimization framework;
- more efforts devoted to multidisciplinary case studies that will serve to test and validate coupling and integration methodologies.

Notes

* In this chapter, the authors made the following contributions: A. Giacomelli – GIS aspects; C. Giupponi – agri-environmental science and decision support; C. Paniconi – hydrology and modelling.

1. The model of Smith *et al.* (1997) takes the form:

$$L_i = \sum_{n=1}^{N} S_{n,i}$$

$$S_{n,i} = \sum_{j \in J(i)} s_{n,j} D_n(Z_j) K(T_{i,j})$$

$$D_n(Z_j) = \beta_n \exp(-\alpha^T Z_j)$$

$$K(T_{i,j}) = \exp(-\delta^T T_{i,j})$$

where L_i is the contaminant transport in stream reach i and $S_{n,i}$ is the contaminant load from source n delivered to reach i from all reaches in subbasin $J(i)$. The N sources for each stream reach i include both point and nonpoint contaminants, and are delivered to the outlet in a two-stage process: land surface to channel network and channel network to outlet at reach i. Thus the second equation contains the contribution to the contaminant mass from source n, $s_{n,j}$, from each reach j of the subbasin, the proportion D_n of this mass that is delivered as a function of land surface characteristics Z_j, and the proportion K transported as a function of channel characteristics $T_{i,j}$. The functional form of the two stages of the delivery process is empirically parameterized as the third and fourth equations, where β_n accounts for sources, α is a vector of delivery coefficients associated with land surface characteristics, and δ is a vector of decay coefficients associated with channel or flow path characteristics. In a demonstration of the interpretive uses of the model, two applications are described. The first is to estimate the proportion of watersheds in the U.S. that meet national clean water standards (for phosphorus, outflow TP concentrations <0.1 mg/l). The second application is to classify watersheds on a region-by-region basis according to predicted total nitrogen yield thresholds of TN $<$ 500 kg/kg^2/yr and TN $<$ 1000 kg/kg^2/yr.
2. Richards' equation may be written as

$$\sigma \frac{\partial \psi}{\partial t} = \nabla \cdot [K_s K_r (\nabla \psi + \eta_z)] + q$$

where $\sigma(\psi)$ is the general storage term or overall storage coefficient, ψ is the pressure head, t is time, ∇ is the gradient operator, K_s is the saturated hydraulic conductivity tensor, $K_r(\psi)$ is the relative hydraulic conductivity, η_z is a vector equal to zero in its x and y components and 1 in its z component, z is the vertical coordinate directed upward, and q is the injected (positive)/extracted (negative) volumetric flow rate (Philip, 1969). The general storage term can be expressed as

$$\sigma = S_w S_s + \phi \frac{dS_w}{d\psi}$$

where $S_w = \theta/\theta_s$, θ is the volumetric moisture content, θ_s is the saturated moisture content (generally equal to the porosity ϕ), and S_s is the specific storage.
3. The equation describing the transport of a reactive contaminant in variably saturated porous media may be written as

$$\phi S_w R_d \left(\frac{\partial c}{\partial t} + \lambda c \right) = \nabla \cdot (D \nabla c) - v \cdot \nabla c + q(c^* - c) + f$$

where R_d is the retardation factor representing adsorption, c is the concentration of the solute, c^* is the decay constant, D is the dispersion tensor, λ is the Darcy velocity vector, c^* is the solute concentration in the injected/extracted fluid, and f is the volumetric rate of injected (positive)/extracted (negative) solute that does not affect the velocity field (Bear, 1979; Huyakorn and Pinder, 1983; Gambolati *et al.*, 1993). The dispersion tensor $D = \phi S_w \tilde{D}$, where $\tilde{D}$ is defined as in Bear (1979), is given by

$$D_{ij} = \phi S_w \tilde{D}_{ij} = \alpha_T |v| \delta_{ij} + (\alpha_L - \alpha_T) \frac{v_i v_j}{|v|} + \phi S_w D_o \tau \delta_{ij} \quad i, j = x, y, z$$

where α_L and α_T are the longitudinal and transverse dispersivity coefficients, respectively, $|v|$ is the magnitude of the velocity vector, δ_{ij} is the Kronecker delta, D_o is the molecular diffusion coefficient, and τ is the tortuosity ($\tau = 1$ is usually assumed).

4. The coupled model of density-dependent variably saturated flow and miscible salt transport can be expressed as (Gambolati *et al.*, 1999)

$$\sigma \frac{\partial \psi}{\partial t} = \nabla \cdot \left[K_s \frac{1 + \varepsilon c}{1 = \varepsilon' c} K_r (\nabla \psi + (1 + \varepsilon c) \eta_z) \right] - \phi S_w \varepsilon \frac{\partial c}{\partial t} + \frac{\rho}{\rho_o} q$$

$$v = -K_s \frac{1 + \varepsilon c}{1 + \varepsilon' c} K_r (\nabla \psi + (1 + \varepsilon c) \eta_z)$$

$$\phi \frac{\partial S_w c}{\partial t} = \nabla \cdot (D \nabla c) - \nabla \cdot (cv) + qc^* + f$$

where all variables are as previously defined except that now the general storage term σ is a function of both pressure head and concentration.

5. Given the intrinsic multidisciplinarity with which GIS has evolved, several definitions, stemming from different perspectives, have been proposed. Generally, we may encounter definitions emphasizing a functional flow (e.g. "a system for capturing, storing, checking, manipulating, analysing and displaying data which are spatially referenced to the Earth") (Department of the Environment, 1987), those following a content approach focusing on data (e.g. an information system that is designed to work with data referenced by spatial or geographic coordinates. In other words, a GIS is both a database system with specific capabilities for spatially-referenced data, as well as a set of operations for working with the data; Star and Estes, 1990), and, lastly, those adding to data and tools a human resources component (e.g. a system of hardware, software, data, people, organizations, and institutional arrangements for collecting, storing, analysing, and disseminating information about areas of the Earth; Dueker and Kjerne, 1989, or organized activity by which people measure and represent geographic phenomena then transform these representations into other forms while interacting with social structures; Chrisman, 1999).

References

Addiscott, T.M., Wagenet, R.J., 1985, Concepts of solute leaching in soils: a review of modelling approaches, Journal of Soil Science, 36, 411–424.

Association for Geographic Information, 1991, GIS Dictionary, A Standards Committee Publication of the Association of Geographic Information, UK. Version 1.1, STA/06/91.

Atkinson, S.F., 1987, Ground water contamination from saltwater intrusion and limitations on agricultural activities, in Fairchild, D.M., ed., Ground Water Quality and Agricultural Practices, Chelsea, MI: Lewis Publisher.

Attia, F.A.R., 1998, The vulnerability of groundwater to pollution and strategies for its protection under the farming conditions in the Mediterranean Countries, Medit, 2, 53–59.

Band, L.E., 1986, Topographic partition of watersheds with digital elevation models, Water Resources Research, 22, 15–24.

Batty, M., Xie, Y., 1994, Modelling inside GIS: Part 1. Model structures, exploratory spatial data analysis and aggregation, International Journal of Geographical Information Systems, 8, 291–307.

Bear, J., 1979, Hydraulics of Groundwater, New York, NY: McGraw-Hill.

Bear, J., Cheng, A.H.-D., Sorek, S., Ouazar, D., Herrera, I., eds., 1999, Seawater Intrusion in Coastal Aquifers – Concepts, Methods and Practices, Dordrecht, The Netherlands: Kluwer Academic.

Bixio, A.C., Orlandini, S., Paniconi, C., Putti, M., 1999, Coupled surface runoff and subsurface flow model for catchment simulations, in Feyen, J., Wiyo, K., eds., Modelling of Transport Processes in Soils at Various Scales in Time and Space, pp. 583–591, Wageningen, The Netherlands: Wageningen Pers.

Bloschl, G., Sivapalan, M., Gupta, V.K., Beven, K.J., 1997, Preface to the special section of papers from the Fourth International Workshop on Scale Problems in Hydrology. Water Resources Research, 33, 2881.

Burrough, P.A., 1986, Principles of GIS for Land Resource Assessment. Oxford: Clarendon Press.

Chrisman, N.R., 1999, What does GIS mean? Transactions in GIS, 3, 175–186.

Collin, J.J., 1995, Some aspects of desertification seen from the groundwater viewpoint, in Fantechi *et al.*, eds., Desertification in a European Context: Physical and Socio-Economic Aspects. Brussels: European Commission, DG12, pp. 111–121.

Crosbie, P., 1996, Object-oriented design of GIS: a new approach to environmental modelling, in Goodchild, M.F. *et al.*, eds., GIS and Environmental Modelling: Progress and Research Issues, GIS World Books, 383–386.

Department of the Environment, 1987, Handling Geographic Information, Technical report, Her Majesty's Stationary Office.

Domenico, P.A., Schwartz, F.W., 1990, Physical and Chemical Hydrogeology. New York, NY: John Wiley and Sons.

Dueker, K., Kjerne, D., 1989, Multipurpose Cadastre: Terms and Definitions. Bethesda, MD, American Congress on Surveying and Mapping.

EC (European Communities), 1991, Council Directive concerning the protection of waters against pollution caused by nitrates from agricultural sources (91/676/EEC), OJ L 375, 31.12.1991; Bull. 12–1991, point 1.2.298.

EC (European Communities), 2000, Council Directive establishing a framework for Community action in the field of water policy. (2000/60/EC), OJ L327, 22.12.2000.

Ehlers, M., Edwards, G., Bedard, Y., 1989, Integration of remote sensing with geographic information systems: a necessary evolution, Photogrammetric Engineering and Remote Sensing, 55, 1619–1627.

Entekhabi, D., 1995, Recent advances in land–atmosphere interaction research, in Pielke, R.A. Sr., Vogel, R.M., eds., U.S. National Report to International Union of Geodesy and Geophysics 1991–1994: Contributions in Hydrology. Washington, DC: American Geophysical Union, pp. 995–1003.

EPA, 1998, BASINS Version 2.0 User's Manual: Better Assessment Science Integrating Point and Nonpoint Sources. Washington: U.S. Environmental Protection Agency.

European Commission, 1998, A Strategic View of GIS Research and Technology Development for Europe. Technical report, EUR 18126.

FAO (Food and Agriculture Organization of the United Nations), 1979, Groundwater pollution: technology, economics and management, Rome, FAO Irr. And Drain. Paper 31.

FAO (Food and Agriculture Organization of the United Nations), 1996, Control of water pollution from agriculture, Rome, FAO Irr. And Drain. Paper 55.

Frind, E.O., 1982, Simulation of long-term transient density-dependent transport in groundwater, Advances in Water Resources, 5, 73–88.

Galeati, G., Gambolati, G., Neuman, S.P., 1992, Coupled and partially coupled Eulerian–Lagrangian model of freshwater–seawater mixing, Water Resources Research, 28, 149–165.

Gallo, C., Paniconi, C., Gambolati, G., 1996, Comparison of solution approaches for the two-domain model of nonequilibrium transport in porous media, Advances in Water Resources, 19, 241–253.

Gambolati, G., Paniconi, C., Putti, M., 1993, Numerical modelling of contaminant transport in groundwater, in Petruzzelli, D., Helfferich, F.G., eds., Migration and Fate of Pollutants in Soils and Subsoils. Volume 32 of NATO ASI Series G: Ecological Sciences, Berlin: Springer-Verlag, pp. 381–410.

Gambolati, G., Putti, M., Paniconi, C., 1996, Projection methods for the finite element solution of the dual-porosity model in variably saturated porous media, in Aral, M.M., ed., Advances in

Groundwater Pollution Control and Remediation. Volume 9 of NATO ASI Series 2: Environment. Dordrecht, the Netherlands: Holland Kluwer Academic, pp. 97–125.

Gambolati, G., Putti, M., Paniconi C., 1999, Three-dimensional model of coupled density-dependent flow and miscible salt transport, in Bear, J., Cheng, A.H.-D., Sorek, S., Ouazar, D., Herrera, I., eds., Seawater Intrusion in Coastal Aquifers – Concepts, Methods and Practices. Dordrecht, The Netherlands: Kluwer Academic, pp. 315–362.

Giardini, L., Giupponi, C., 1995, Modelli agricoli e impatto ambientale. Milan: Franco Angeli.

Giupponi, C., 1995, Modelling agriculture and the environment: crop production and non-point source pollution, European Journal of Agronomy, 4, 403–412.

Giupponi, C., 1998, Environmental evaluation of alternative cropping systems with impact indices of pollution, European Journal of Agronomy, 8, 71–82.

Giupponi, C., Ghetti, P.F., 1996, Valutazione dell'impatto sulla qualità delle acque di sistemi agricoli alternativi nel bacino scolante in Laguna di Venezia, VII Congresso Nazionale SitE, Naples.

Giupponi, C., Rosato, P., 1995, Simulating impacts of agricultural policy on nitrogen losses from a watershed in northern Italy, Environment International, 21, 577–582.

Giupponi, C., Rosato, P., 1998, A farm multicriteria analysis model for the economic and environmental evaluation of agricultural land use, in Beinat, E., Nijkamp, P., eds., Multicriteria Evaluation in Land Use Management. Dordrecht, the Netherlands: Kluwer Academic, pp. 115–138.

Giupponi, C., Rosato, P., 1999, Agricultural land use changes and water quality: a case study in the watershed of the Lagoon of Venice, Water Science and Technology, 39, 135–148.

Grayson, R.B., Blöschl, G., Barling, R. D., Moore, I.D., 1993, Process, scale and constraints to hydrological modelling in GIS. In HydroGIS 93: Application of Geographic Information Systems in Hydrology and Water Resources. Vienna: IAHS, pp. 83–92.

Grenon, M., Batisse, M., 1989, Futures for the Mediterranean Basin. The Blue Plan. New York: Oxford University Press.

Hassler, B., Lecca, G., Paniconi, C., 1999, Problems and features of saltwater upconing due to well pumping, CRS4 Technical Report, Dec. 1999, Cagliari, Italy.

Henry, H.R., 1965, Effects of dispersion on salt encroachment in coastal aquifers, in Sea Water in Coastal Aquifers. US Geological Survey Water Supply Paper, No. 1613-C, pp. 70–84.

Holden, P.W., 1986, Pesticides and Groundwater Quality. Washington, DC: National Academy Press.

Huyakorn, P.S., Pinder, G.F., 1983, Computational Methods in Subsurface Flow. New York, NY: Academic Press.

Huyakorn, P.S., Andersen, P.F., Mercer, J.W., White, H.O., 1987, Saltwater intrusion in aquifers: development and testing of a three-dimensional finite element model, Water Resources Research, 23, 293–312.

Jenson, S.K., Domingue, J.O., 1988, Extracting topographic structure from digital elevation data for geographic information system analysis, Photogrammetric Engineering and Remote Sensing, 54, 1593–1600.

King, J.L., Kraemer, K.L., 1993, Models, facts, and the policy process: the political ecology of estimated truth, in Goodchild, M.F., Parks, B.O., Steyaert, L.T., eds., Environmental Modelling with GIS. Oxford University Press, pp. 353–360.

Kinzelbach, W., Schafer, W., 1993, Modelling and design of in-situ bioremediation measures, in Groundwater Quality Management. IAHS Publ. No.220, pp. 399–419.

Kite, G.W., 1995, Scaling of input data for macroscale hydrologic modelling. Water Resources Research, 31, 2769–2781.

Kolditz, O., Ratke, R., Diersch, H.-J.G., Zielke, W., 1998, Coupled groundwater flow and transport: 1. Verification of variable density flow and transport models. Advances in Water Resources, 21, 27–46.

Kuniansky E.L., Lowther, R.A., 1993, Finite-element mesh generation from mappable features. International Journal of Geographical Information Systems, 7, 395–405.

Leipnik, M.R., Kemp, K., Loaiciga, H.A., 1993, Implementation of GIS for water resources planning and management. Journal of Water Resource Planning and Management, 119, 184–205.

Livingstone, D., Raper, J., 1994, Modelling environmental systems with GIS: theoretical barriers to progress, in Worboys, M.F., ed., Innovations in GIS. London: Taylor & Francis, pp. 229–240.

Maidment, D.R., 1993, GIS and hydrologic modelling, in Goodchild, M.F., Parks, B.O., Steyaert, L.T., eds., Environmental Modelling with GIS. New York, NY: Oxford University Press, pp. 147–167.

Maidment, D.R., 1996, Environmental modelling within GIS, in Goodchild, M.F. *et al.*, eds., GIS and Environmental Modelling: Progress and Research Issues. GIS World Books, pp. 315–323.

Mangold, D.C., Tsang, C.F., 1991, A summary of subsurface hydrological and hydrochemical models. Revues in Geophysics, 29, 51–79.

Merchant, J.W., 1994, GIS-based groundwater pollution hazard assessment: a critical review of the DRASTIC model, Photogrammetric Engineering and Remote Sensing, 60, 1117–1127.

Miller, M.R., Brown, P.L., Donovan, J.J., Bergatino, R.N., Sonderegger, J.L., Schmidt, F.A., 1981, Saline seep development in the North American Great Plains – hydrological aspects, in Holmes, J.W., Talsma, T., eds., Land and Stream Salinity. Amsterdam: Elsevier, pp. 115–14.

Moore, I.D., Turner, A.K., Wilson, J.P., Jenson, S.K., Band, L.E., 1993, GIS and land–surface–subsurface process modelling, in Goodchild, M.F., Parks, B.O., Steyaert, L.T., eds., Environmental Modelling with GIS. New York, NY: Oxford University Press, pp. 196–230.

Novotny, V., Olem, H., 1994, Water quality. Prevention, identification and management of Diffuse pollution. New York (USA): Van Nostrand Reinold, 1050 p.

Nyerges, T.L., 1994, Understanding the scope of GIS: its relationship to environmental modelling, in Worboys, M.F., ed., Innovations in GIS. Taylor & Francis, pp. 75–93.

OECD (Organisation for Economic Cooperation and Development), 1994, Environmental Indicators – OECD Core Set. Paris: OECD.

OGC Technical Committee, 1998, The OpenGIS Guide. Wayland, MA: the OpenGIS Consortium.

Paniconi, C., Putti, M., 1994, A comparison of Picard and Newton iteration in the numerical solution of multidimensional variably saturated flow problems. Water Resources Research, 30, 3357–3374.

Paniconi, C., Kleinfeldt, S., Deckmyn, J., Giacomelli, A., 1999, Integrating GIS and data visualization tools for distributed hydrologic modelling. Transactions in GIS, 3, 97–118.

Parker, J.C., 1989, Multiphase flow and transport in porous media. Revues in Geophysics, 27, 311–328.

Peirce, M., 1998, Computer-based models in integrated environmental assessment, Technical Report AEAT-1987, AEA Technology.

Perez-Trejo, F., 1992, Desertification and land degradation in the European Mediterranean, EPOCH Programme, EC-DG12, EUR 14850 EN.

Philip, J.R., 1996, Theory of infiltration. Advances in Hydroscience, 5, 215–296.

Putti, M., Paniconi, C., 1995, Picard and Newton linearization for the coupled model of saltwater intrusion in aquifers. Advances in Water Resources, 18, 159–170.

Raper, J., Livingstone, D., 1996, High-level coupling of GIS and environmental process modelling, in Goodchild, M.F. *et al.*, eds., GIS and Environmental Modelling: Progress and Research Issues. GIS World Books, pp. 387–390.

Renard, K.G., Rawls, W.J., Fogel, M.M., 1982, Currently available models, in Haan, C.T., Johnson, H.P., Brakensiek, D.L., eds., Hydrologic Modelling of Small Watersheds. St. Joseph, MI: American Society Agricultural Engineering, pp. 507–522.

Richards, J.A., 1986, Remote Sensing Digital Image Analysis, An Introduction. New York: Springler-Verlag.

Ritchie, J.T., 1987, Using crop models as a decision support system to reduce nitrate leaching, in D'Itri, F.M., Wolfson, L.G., eds., Rural Groundwater Contamination. Lewis Publisher, pp. 179–192.

Schincariol, R.A., Schwartz, F.W., Mendoza, C.A., 1994, On the generation of instabilities in variable density flow. Water Resources Research, 30, 913–927.

Ségol, G., 1993, Classic Groundwater Simulations: Proving and Improving Numerical Models. Englewood Cliffs, NJ: PTR Prentice Hall.
Sherif, M.M., Singh, V.P., 1996, Saltwater intrusion, in Singh, V.P., ed., Hydrology of Disasters. Dordrecht, the Netherlands: Kluwer Academic, pp. 269–316.
Simmons, C.T., Narayan, K.A., Wooding, R.A., 1999, On a test case for density-dependent groundwater flow and solute transport models: the salt lake problem. Water Resources Research, 35, 3607–3620.
Simonovic, S.P., 1997, Risk in sustainable water resources management, in Sustainability of Water Resources Under Increasing Uncertainty. IAHS Publ. 240, pp. 3–17.
Smith, R.A., Schwarz, G.E., Alexander, R.B., 1997, Regional interpretation of water-quality monitoring data. Water Resources Research, 33, 2781–2798.
Star, J., Estes, J., 1990, Geographic Information Systems: An Introduction. Englewood Cliffs, NJ: Prentice Hall.
Suarez, D.L., 1989, Impacts of agricultural practices on groundwater salinity. Agriculture, Ecosystems and Environment, 26, 215–227.
Taigbenu, A.E., Liggett, J.A., Cheng, A.H.-D., 1984, Boundary integral solution to seawater intrusion into coastal aquifers. Water Resources Research, 20, 1150–1158.
Umali, D.L., 1993, Irrigation induced salinity – a growing problem for development and the environment. World Bank Technical Paper, no. 215.
UNITAR (United Nations Institute for Training and Research), 1995, GIS and decision making. Geneva, Switzerland: UNITAR, 128 pp.
United Nations Environment Program (UNEP), 1999, A survey of spatial data handling technologies 1997, environmental and assessment technical report 13, Technical report, United Nations Environment Programme, Sioux Falls, SD: USA Global Resources Information Database. Online. Available: http://grid2.cr.usgs.gov/survey/97survey/survey.html.
van Genuchten, M.T., Nielsen D.R., 1985, On describing and predicting the hydraulic properties of unsaturated soils, Annals of Geophysics, 3, 615–628.
van Genuchten, M.T., Simunek, J., 1996, Evaluation of pollutant transport in the unsaturated zone, in Rijtema, P.E., Elias, V., eds., Regional Approaches to Water Pollution in the Environment. Dordrecht, the Netherlands: Kluwer Academic, pp. 139–172.
van Keulen, H., Wolf, J., 1986, Modelling of agricultural production: weather, soils and crops. Wageningen (NL): Pudoc.
Voss, C.I., Souza, W.R., 1987, Variable density flow and solute transport simulation of regional aquifers containing a narrow freshwater–saltwater transition zone. Water Resources Research, 23, 1851–1866.
Welsh, D.J., 1991, Riparian forest buffers. Function and design for protection and enhancement of water resources. Radnor, PA: USDA.
Wilson, J.P., 1996, GIS-based land surface/subsurface modelling: new potential for new models?, in Proceedings Third International NCGIA Conference on Integrating GIS and Environmental Modelling, CD-ROM, Santa Barbara, CA: NCGIA.
Yurdisef, M.A., Jamieson, D.G., 1997, Development of a water resources planning model which includes environmental considerations, in Sustainability of Water Resources Under Increasing Uncertainty. IAHS Publ. 240, pp. 419–425.
Zalidis, G., 1998, Management of river water for irrigation to mitigate soil salinisation on a coastal watershed. J. Environmental Management, 54, 161–167.
Zienkiewicz, O.C., 1986, The Finite Element Method. New York, NY: McGraw-Hill.
Zimmerman, D.A., de Marsily, G., Gotway, C.A., Marietta, M.G., Axness, C.L., Beauheim, R.L. *et al.*, 1998, A comparison of seven geostatistically based inverse approaches to estimate transmissivities for modelling advective transport by groundwater flow. Water Resources Research, 34, 1373–1413.

CHAPTER 4

Issues in the Valuation of Groundwater Benefits

Nir Becker and Stefania Tonin

4.1. Introduction

Why do we need to put special effort into finding an aquifer's value?[1] Estimating the value of a aquifer is useful for both decisions concerning resource allocation among different users as well as long term decisions with respect to investments in restoration, conservation prevention and developing alternative water supplies. The value of an aquifer and the long-run analysis of groundwater use are strongly correlated to the problem of sustainability. It can be claimed that a society is not moving along a sustainable water resources management path if one of the following conditions hold:

(a) the decline in the quantity and quality of available freshwater is not compensated for by more effective and efficient water services;
(b) the costs of water services required to relax physical resources limits are subject to continuous increase over time;
(c) there are political or social constraints which impede adoption of the institution reforms required to make freshwater resources allocation less socially inefficient.

Theoretically, if water from a given aquifer were traded within a marketing system, there would not be any great difficulty in estimating its value. However, in practice, things are not so simple. Groundwater resources are usually characterized by free access. It is therefore likely that there will be some difference between the private value of groundwater and its social value. Prices, which should reflect opportunity costs are not as efficient as prices in a well-functioning market. Moreover, as will be discussed later on, some of the services that are provided by the aquifer are not priced at all. There is clearly some need for governmental intervention. This intervention should be based on the total economic value of the aquifer, and policy-makers should be aware of how different policies could affect this value. Policies should of course be targeted in order to maximize the social value of the aquifer: if the linkage between policy and value is uncertain, it is impossible to carry out such a calculation.

The main aim of this chapter is to deal with the different approaches that

C. Dosi (ed.), Agricultural Use of Groundwater, 77–106.

should be taken in order to estimate the total economic value of an aquifer. It should be viewed as the first step towards establishing a policy agenda. Such a policy agenda is not discussed in this chapter but is not less important.

The chapter is organized as follows. Section 2 is devoted to a description of the rationale behind the concept of total economic value (TEV) and a suggested taxonomy for the different services a given aquifer can provide. Section 3 describes the different approaches used in environmental economics to value these services. Some of them my undergo through a market process and some not. We provide a guiding list to enable decision makers to get an initial idea of what kind of research should be done with respect to a given aquifer. We will then mention some case studies at the hand of a literature review. This will show how difficult it is to get the TEV of an aquifer. The considerable effort involved in finding this is related to the fact that the various groundwater aquifers provide different services. Section 4 provides a summary and recommendations for steps which should be taken as well as the particular research efforts needed with respect to valuation issues and policy formation.

4.2. Groundwater valuation – rationale and taxonomy

4.2.1. *The rationale for valuation*

Goods that are characterized as private goods can be traded in a market system. The interaction between supply and demand forces creates an equilibrium price. If property rights are well defined, individuals can buy and sell according to that market price. The solution derived is then an efficient one. There are, however, goods or services that do not belong to anyone in particular. In other words, they are common property resources (CPR). For such resources, the term equilibrium price has no meaning. This is because a potential buyer would find out that he can get the goods without paying and a potential seller would therefore not find matching buyers.

However, the fact that these goods do not have a market price does not mean that they do not have a value of their own. They certainly do, but finding it is a more complicated task. In the case of groundwater resources, their value can be broken down into both an out of pocket cost and an opportunity cost. The first term refers to the energy cost related to pumping and the usual maintenance cost. The second term refers to what could have been done with the extracted unit of water had it stayed within the aquifer and not been extracted.

As will be emphasized later on, groundwater resources provide various services. Some of them can be considered as being related to the opportunity cost, and do not have a market price associated with them.

The valuation techniques summarized in this chapter can help decision-makers in the water policy area in different ways. Most important of these is

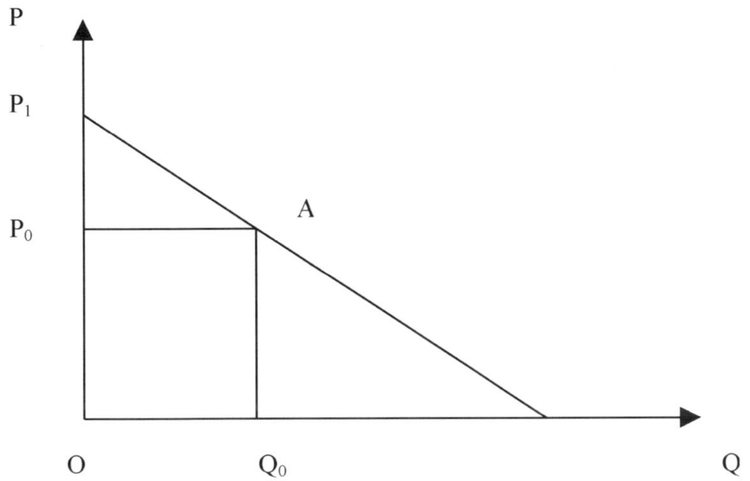

Figure 4.1. The demand curve.

that they can help policy makers understand how they can intervene in order to maximize the value of the given aquifer. Alternatively, they can help them understand how specific policies affect each aquifer value and its total value. In this regard, by showing the implications of a specific policy rather then asking what the best policy is, the concepts suggested in this chapter can serve a crystal ball function.

4.2.2. *The linkage between prices and values*

If we start from a usual demand function, then given a price level, we can find out the quantity consumed at this price. However, for our purposes, it would be better to start from the quantity as given and then ask what the vertical difference to the (inverse) demand curve tells us. The answer to that question is that it tells us the marginal willingness to pay for that amount. Looking at Figure 4.1, we can see that if the marginal willingness to pay for the Q_0 unit is less then P_0 then one would not buy that unit. Hence, it would not be consistent with the demand function as revealed by the consumer. The same argument, only in reverse, applies with respect to a potential willingness to pay which is higher than P_0.

We can then conclude that the demand curve for a given commodity is its marginal willingness to pay curve. If we have a marginal willingness to pay, then from it we can derive the total willingness to pay by finding the area under the demand curve, as Figure 4.1 shows. This is correct for every quantity. For example, if we wish to estimate the total willingness to pay for Q_0 units of that good (water, say), then this is given by the area P_1AQ_0O. This is the linkage between prices and benefits. Knowing prices enables us to deduce the

Table 4.1. Groundwater services

Extractive value (flow)	In situ value
Agricultural water use	Stock value
Residential water use	Avoiding sea water intrusion
Industrial water use	Buffer value
	Option value
	Subsidence avoidance
	Recreation
	Ecological values

demand curve and by doing so, this enables us to deduce the total willingness to pay. The total willingness to pay is the benefit provided by the asset, or as we called it, the value of the asset.

In cases where market prices are given and a demand curve can be traced, then the question of valuing the given asset is relatively straightforward. However, our aim is to construct a demand function for each value, even if there is no market value associated with it. This will be illustrated later on, with reference to groundwater. However, before turning to the techniques used to measure the value of an aquifer, we will firstly look at the issue of identifying the different services that an aquifer can provide.

4.2.3. *A taxonomy for groundwater services*

There is no unique way of listing the different values of a groundwater aquifer. However, we suggest using two distinct values, and within these two categories, divide each one of them into sub-categories. The two main categories we suggest are extractive (flow) values and in situ values. A list of the sub-categories is presented in Table 4.1.

The aquifer services are divided into flow and in situ in the table. Within the flow values are agricultural, residential and industrial water uses. These values are associated with the amount of water extracted from the aquifer. It should be noted that here, the demand for each value is dependent on the quality of the water. That makes the analysis even more complex, since taking water from the aquifer not only reduces the available use for other alternatives but may cause degradation in the water quality as well. Thus, we not only have a shift along the demand curve for other uses, but the demand curve itself shifts. Unless this is taken into account, the benefit valuation will be misleading.

For a given quality, however, we can list three major extractive uses for the water in a given aquifer, namely agricultural, residential and industrial.

- Agricultural water use: This is simply a flow of water emanating from the aquifer which is used either to irrigate crops or for livestock watering. The cost of abstraction as well as the quality of the water obviously affects the demand for water. As will be seen later on, only part of the cost is borne by the farmer himself, while the rest is a classical externality case. So even here, where it is a matter of a classical example of market good (agricultural commodities), there are still imperfections and a need to estimate the demand for water not through the market, as will be explained later on.

- Residential water use: By this is meant residential water uses of all kinds, though it will mainly consist of domestic and water that is lent out. The major concern here is quality. However, quantity issues are also important, especially in dry seasons where water could be saved by not being used for outside uses (gardens, swimming pools, etc.). These two kinds of concern will be outlined later on.
- Industrial water use: This category will mainly consist of water that has originated in a groundwater aquifer and is used for cooling, hydropower electric production, washing and also for the beverage industry. Conditions of quality are not as important as in residential water use. Moreover, responsiveness to price varies considerably in the short and long-run.

We now turn to the other type of services, namely, the in situ values. These values are associated with the opportunity cost of taking a unit of water now. It should be noted that in a steady state situation, the average withdrawal should be equal to the long-run average recharge of water. Policy makers sometimes ignore the simple truth of water in equals water out. Thus, the important question is not how much water originating from the aquifer we should use. The answer to that question is quite simple. The relevant question should be what is the best (or optimal) water level of the aquifer and how changes in this level affect the different aquifer values.

We recognize several in situ aquifer values (see Table 4.1). We will later describe different approaches to measuring these values and some efforts made in estimating them. However, we will firstly try to explain what each one of them means.

Stock value

The stock value is related to the aquifer's water table. If pumping the same amount of water from a higher water table entails a smaller energy cost, then the water stock plays a role in determining the value of the aquifer. Although we recognize that the amount of water that is used would be the same in a steady state (and equal to the recharge), we would prefer to pump it from a higher level. The opportunity cost of taking one unit of water today is the discounted present value of the cost incremental to future pumping from a lower water table. This is called the dynamic shadow price of the aquifer.

The benefits from avoiding sea water intrusion

This value is mainly applicable to coastal aquifers. If the water level is decreasing, there is an increasing chance that sea water will start diffusing into specific part of the aquifers (see Chapter 2, this volume). The problem is associated with uncertainty because this critical level is not known for sure. Once it happens, it is not reversible in the foreseeable future. Therefore, keeping water tables high serves as an insurance against such damage.

The buffer value

This value is associated with uncertainty with regard to future availability of water supplies. The uncertainty here is with respect to the aquifer's recharge rate or the supply of surface water that serves the same area. If water users are risk-neutral and have only been responding to the mean recharge, then there would not be any buffer value to the aquifer. However, when individual users are risk averse, then uncertainty with respect to the recharge rate or alternative water supply sources has negative implications for them. In such cases, groundwater has a buffer value which is equal to the difference between the maximal value of the aquifer under uncertainty and its maximal value under certainty where the recharge rate is stabilized at its mean (Tsur and Graham-Tomasi, 1991).

The option value

This value is also connected to uncertainty. Here, the uncertainty is about future preferences. That is, suppose, for example, that we anticipate using the aquifer in the future but are not sure about this. Over-exploiting the aquifer will prevent us from even considering the option of using it in the future. This is correct only if the damage is irreversible to some extent. We can then define the option value as the amount of money people would be willing to pay in order to avoid the risk of that scenario.

Benefits from subsidence avoidance

In many areas, groundwater stocks contribute to sustain underground water/soil structure so as to prevent land subsidence. Land subsidence generally occurs when aquifer pressure levels are significantly lowered in basins where the substrate is primarily fine-grained material such as clays and silts, which are more compressible than more rigid coarse grains such as sand and limestone and sandstone formations. Subsidence caused by the consolidation of fine-grained material cannot be reversed by artificially injecting additional water into the formation. Subsidence is reversible only in aquifers usually dominated by sands, gravels, or sandstone, which can accept the additional fluids (National Research Council, 1997). On all continents, subsidence has been observed in certain areas where groundwater is over-exploited for human use, or where groundwater is extracted for underground access, such as in mines. In shallow coastal areas, subsidence may cause flooding, as the land literally sinks into the sea. Water stored in aquifers can also help to control flooding and erosion by providing a medium for absorbing surface water runoff. The underground water/soil structure of an aquifer also represents a medium for the absorption, transport, and dilution of wastes and other by-products of human economic activity.

Recreational values
This is important when the groundwater is connected to surface water. When the water table is declining, slow outflow in the connected streams is one of the consequences. This, in turn, affects an entire range of recreational uses (swimming, fishing, and boating, for example).

Ecological performance
Again, this is important when ground and surface water are connected. Reducing the outflow due to decreased water table in the adjacent aquifer may prove to be harmful to wetland and the wide variety of species that use it as a habitat.

We have thus listed 10 different potential groundwater services: three extractive values and seven in situ values. There is clearly no clear-cut distinction between the different values. For example, one can see recreational and ecological values as being extractive in nature. The important thing, however, is to list them all and find their relative importance. This will, of course, differ on a case-by-case basis.

Finally, when adding the different values of the aquifer in order to get a total economic value, care should be taken with respect to the time dimension. Some of the values are given as yearly values while others are given as present values. For example, the stock value represents a form of the discounted sum of future increased costs. Care should therefore be taken in adding numbers of different dimensions.

4.3. Alternative approaches to groundwater valuation

4.3.1. *General framework*

In all the cases listed below, we use the willingness to pay criterion as a measure of the benefit or value. In the case of a cost–benefit analysis of a regulatory decision (e.g. regulation of groundwater abstractions, adoption of conservation measures aimed at reducing groundwater pollution), the difference in value between doing with or without the public programme will serve as the social benefit provided by the regulation.[2]

We refer to three categories of methods to value groundwater services: methods relying on market information, methods relying on surrogate (proxy) markets and methods relying on hypothetical markets. Relying on market information should be used whenever markets to the service exist and are well functioning. For example, a change in the value of agricultural production whenever there is a market for agricultural products falls into this category.

Table 4.2. General framework

Groundwater services	Applicable valuation method
Extractive	
• Agricultural water use	
Quantity	linear programming method, farm budget residual approach, relationship between water applications and crop yields
Quality	contingent valuation, crop response model
• Residential water use	
Quantity	demand function
Quality	contingent valuation, hedonic price method, avoidance cost, cost of illness, benefit transfer
• Industrial water use	
Quantity	demand function
Quality	contingent valuation
In situ	
• Stock value (quantity)	energy price, dynamic optimization
• Avoiding sea water intrusion (quality)	dynamic optimization, averting cost method
• Buffer value	dynamic optimization, contingent valuation
• Option value	contingent valuation, dynamic optimization
• Subsidence avoidance	hedonic price method, contingent valuation
• Recreation	travel cost method, contingent valuation
• Ecological values	contingent valuation

However, even here, care must be taken to implement market prices as a marginal benefit proxy if there is government intervention within the crop market itself (e.g. subsidies). The other cases that fall in this category will be listed below.

The other two categories cover cases in which the specific service is not traded in a market, or alternatively, market prices do not reflect the true social cost. Here, the first approach is one which relies on proxy markets. Here we have goods without a market, though there is a nearby market that is in some way correlated with the absent market. These indirect approaches include, for example, the travel cost method to estimate the value of recreational areas, and hedonic price methods which relate property value changes to different environmental conditions.

On the other hand, the direct methods are used when there are no surrogate markets or these markets do not function well. In such cases, surveys done on the basis of simply asking people questions regarding their willingness to pay for different attributes can be used. As will be seen later, these approaches are the only ones that can estimate some groundwater values. However, particular care should be taken in interpreting results based on hypothetical questions about hypothetical scenarios.

Our general framework is presented in Table 4.2. As can be seen from the table, some of the services have two dimensions, namely quantity and quality. Both should be taken into account.

4.3.2. *Methods based on market prices*

We have to distinguish between water as an intermediate good and water as a final good. The former use applies mainly to agricultural and industrial water uses, while the later applies mainly to residential water uses. We will start by describing the intermediate case and then the final good case. However, we will first describe a method that could be used in both intermediate as well as final good cases.

Water markets

Here, we simply rely on buyer/seller transactions involving an exchange of money for water. If these markets are well functioning, we can find out the true marginal willingness to pay, which is the equilibrium price. The problem is that water markets are more the exemption then the norm. Therefore, whenever policy makers can, they should try to create water markets which are self-revealing with respect to the market price. There are some water markets evolving in southwestern USA (Saliba and Bush, 1987) and in other places in the world (Easter and Hearne, 1995; Rosegrant and Binswanger, 1994). However, even if markets do not exist, they can be hypothetically created in order to find out the equilibrium price of water (Becker, 1995).

Water as an intermediate good

With this method, we can infer the value of the water from those goods for which water serves as an input. According, the value of the agricultural crop should serve as the basis for calculating the value of the groundwater, for example. The best method for estimating this value is by mathematical programming. Using these models allows the planner not only to know what is the best water allocation but also to estimate the value of the water in the groundwater (which is actually what we are after).

An objective function and the constraints will need to be specified:

$$\text{Max. } \Pi = f(\Sigma_i \, VMPW_i \cdot W - \Sigma_i \, TCC_{ij} \cdot W_{ij} \quad (1)$$

S.T.

$$\Sigma_i \, W_{ij} < W_j \quad (2)$$

....

where Π stands for the total profit for the region, $VMPW_i$ stands for the value of the marginal product of water for crop i, W_i stands for the amount of water allocated to crop i, TCW_{ij} stands for the cost of putting water originating from groundwater source j on crop i, W_{ij} stands for the amount of water delivered from groundwater source j to crop i, W_j stands for the total amount of water available in groundwater *j*, ... stands for other technological and institutional constraints.

The results of this programming model yield the optimal allocation of water from different sources to one region with a vector of agricultural markets activities. The model can be extended to include several demand regions by adding another summation element to represent the entire spectrum of demand regions of which region *i* is only one of them.

The method for calculating the VMP for each crop is called the residual return for water. Here, we have to assume that there are competitive markets in the other production factors, which is not such an unreasonable assumption. We also have to determine the competitive price for the commodities themselves, either by direct observation or by a complementary research that deletes all market interventions. In that case, we will obtain the following:

$$P_i = (PT*QT) + (PL*QL) + PK*QK) + (VMPW_i*QW) \qquad (3)$$

where P_i stands for the price of crop *i*, P_j $(j = T, L, K)$ stands for the prices of land and labour, Q_j $(j = T, L, K, W)$ stands for the quantities of the above production factors, $WMPW_i$ stands for the value of marginal product in crop *i*.

Assuming that all other variables except $VMPW_i$ are known, we can solve the last equation for the value of marginal product of water. The data for this calculation is usually obtained from farm accounting data. To further simplify things, we usually normalize one factor, mainly land. That is, all the relevant data (crop price, amount of capital, labour and water, for example) are given per hectare.

Another sector for which water serves as an intermediate good is the industrial sector. In such cases, a value is assigned either by calculating the demand function for water or by calculating the replacement cost. The demand for water in the industrial sector is relatively fixed in the short-run (almost zero) but becomes more flexible in the long-run. Thus, for policy purposes, it is important to estimate the elasticity of demand in that sector for the short and long-run and also the switching point between the two time spans. The main difficulty with estimating the demand function for industrial uses is the lack of sufficient variation in prices. A major part of the water-using industries uses self-supply water, which turns out not to be separable from the total firm cost.

Having obtained the demand function, the next step is to calculate the value of the water for the specific industry. This is done, as explained in Section 4.2.2, by calculating the area under the demand curve over a relevant range. Say we have a constant elasticity of demand (which is consistent with a logarithmic demand function where the price coefficient represents the elasticity). Then if we denote elasticity by γ, we can write the area under the demand curve for a change in the quantity from Q_0 to Q_1 in the following way:

$$B = \{(p_0 Q^{1/\gamma})/(\gamma - 1/\gamma)\}\{Q_0/Q_0^{1/\gamma} - Q_1/Q_1^{1/\gamma}\} \qquad (4)$$

After deducting the payment for that water, the value of the water for the industry can be obtained. It is also quite useful to obtain the value per unit of water, by dividing the net value by the amount of water consumed. Finally, leakage and other water losses in the system should be taken into account by calculating the value on water that was actually received by the consumer.

The method of the replacement cost is an alternative method. This actually assigns a value to the water based on the difference between the cost of production with water as a factor of production and the next best alternative. For example, in hydroelectric power generation, that difference can be estimated per kWh.

Water as a final good

Research in this area is aimed particularly at the municipal water sector. Here, the main component is residential water demand, which refers to all water uses inside or outside the household. The remainder covers such uses as public recreational facilities, schools etc. The theory of water demand for residential use is based on consumer rather than on producer theory, as was the case when water was an intermediate good. The demand function for water is derived from a utility function of the household. The technique for finding the value of the water to the final consumer is the same as for the industrial and agricultural water uses, i.e. finding the area under the demand curve. The difference lies in specifying the demand function. Demand functions for water are usually specified as being determined by the price of water, price of related goods, income, weather, and other regulatory policies (such as conservation).

There are several problems associated with the econometric estimation of residential water demand. The first problem has to do with the lack of price variation, especially with regard to the higher price. The other problem is closely associated with the price mechanism. Residential water is often priced by block rates, usually increasing. This causes two problems. The first is the price perception and the other one is the simultaneity problem. For block rate pricing, price perception should be defined. This is because it is not so obvious that consumers respond to the marginal price. It would appear that they mainly react to the average lagged price one billing period earlier. A regression equation that does not take this into account is likely to yield a biased estimate. This is usually taken into account by specifying a perception variable which captures the difference between what the consumer would actually pay if a marginal price were charged for the entire amount of water consumed relative to what he had paid in reality. The second problem is simultaneity. This mainly occurs because the price of water affects the quantity consumed, but the quantity consumed also affects the price level because of the block rate pricing. Thus, quantity is located in both sides of the demand equation.

So far we have briefly described methods that rely upon markets that exist in reality. A lot of groundwater services, however, are characterized by the absence of markets associated with them. We therefore need a tool kit for calculating these values. As explained before, we will divide these value estimates into two main categories: those that rely on proxy markets and those that rely on direct valuations through surveys and questioners.

4.3.3. *Application of methods based on market price (agricultural, industrial and residential water uses)*

Modelling agricultural water demand is important in that the value of the irrigated water (as reflected by the area under the derived demand curve) can be obtained.

According to Letey (1991), the Von Liebig production function, which assumes a fixed water use per acre for a given crop at a given location, holds at the field level: the estimation is performed by determining land allocation among crops. A sample of such studies can be found in Becker (1995) for Israel, Howeitt (1995) for California and McCarl (1982) for Texas.

The resulting shadow prices for water and land had policy implications. For example, in the case of Israel, the shadow prices for the two main aquifers were about 6 cents/m^3. This could be helpful in designing the impact of water markets or a new water pricing policy. If the total amount of water in the aquifer is known, it can also give information regarding its value. In another study (Becker *et al.*, 1996), the value of Israeli mountain aquifer was estimated in that way. Since this aquifer has to somehow be divided between Israel and the Palestinian authorities, this value can be useful in determining different allocation mechanisms. Finally, it serves as a signal for backstop technologies (e.g. desalination). In the Becker *et al.* study, without optimal allocation, the shadow prices for the Palestinians were found to be \$1/m^3, which is higher than desalination (about 80 cents/m^3). However, the equilibrium price in the optimal allocation was found to be 46 cents/m^3 (including all uses; not only agriculture). Thus, desalination would prove to be a bad solution.

In another study, Sunding *et al.* (1995) found differences in the marginal product of water in the range \$12–60 in the Californian Central Valley project. These different shadow prices and marginal value of water can be justified only for different conveyance costs; thus, given this situation, the value of water is misleading.

The programming models have proved to be particularly helpful since in most cases, water pricing does not represent marginal values. By using these programming models, one can find the shadow price of water for different amounts of water. These shadow prices, together with their matching quantities, could be used to trace the derived demand function for water, as was done in the three studies mentioned above. Boggess *et al.* (1993) found that, in general, the elasticity of water varies: as the price of water increases (corresponding to lower water use levels) demand becomes more inelastic.

Moore and Dinar (1995) took another approach. In their study, they used econometric techniques to estimate water allocation among crops. This, in turn, was used to assess water demand. They found that the implicit shadow prices were much higher than market prices. The correct demand function for water was then derived from the quantity–shadow price relationship. Finally, the water–crop production function can be used to demonstrate the impact of changing input level (water, say) on the output. While this approach is more accurate then the previous ones, its limitations are the large cost involved in conducting such studies (mainly based on field experiments), and the limited application that they have (usually to a very specific site). Dinar and Letey (1994) provide an example of such an analysis while Boggess *et al.* (1993) provide a literature review.

Considering the quality aspects, dissolved mineral salts as well as seawater intrusion are the major water pollutants affecting irrigated crops. They are mainly chlorides, sulphates and nitrates. When water is applied for irrigation, these minerals stay in the irrigated soil, and in turn, crop productivity is adversely affected. The general approach is to add all minerals into one measure called total dissolved solid (TDS). Salinity affects different crops to different extents. However, the least sensitive crop is usually also the less valuable. In order to avoid crop damage from these salts, farmers apply water in excess of the crop needs mainly to drive the salts down below the root zone of the plant. However, by doing so, they affect the groundwater and subsequent users of the water.

The benefit of salinity abatement can be derived using a similar approach to that for estimating the change in the quantity of water. A reduction in the salinity brings about an increase in the net income, which is the perceived benefit of salinity abatement. Letey (1991) is an example of a study that provides a production function which reflects the change in yields due to reduced salinity.

However, attention should also be paid to farmers' response to changing conditions of salinity. This may take on two main forms: moving into more salt-resistant crops (which are, however, less profitable) or shifting to high efficiency irrigation methods such as sprinklers and drip irrigation (which are more expensive). This could be captured by a mathematical programming that reflects the irrigator response to salinity in terms of crop mixture as well as the irrigation technology. An example of such approach is presented in Booker and Young (1994).

As in agriculture, water for industrial use constitutes an intermediate good. Therefore, the derived demand for water in that sector will be a function of the price of water, price of other inputs and price of the final output. When price of water changes, two factors will affect its use: the elasticity of substitution between water and other inputs and its own price elasticity. Kindler and Russel (1984) made a survey of the issues in modelling industrial water demand. They identified two main approaches towards estimating the value of water: using econometric techniques or mathematical modelling.

With regard to the econometric approach, the main problem for estimating the demand function is that the number of cases in which water is volumetrically priced are rather limited. The two main studies using this approach are those by Babin *et al.* (1982) and Renzetti (1992). In the first study, a Cobb–Douglas production function was assumed. This implies, however, a constant unitary demand elasticity for water. The other study allowed for variation in the elasticity. The study was done in respect of industrial water users in Canada. Renzetti reports an average price elasticity of −0.38. Another important result of this study was a realization of the importance of substitutability among inputs. It confirms the fact that recirculation of water is a substitute for both water intake and water discharge. Thus, firms are likely to reduce water intake and increase recirculation. Finally, it was found that there is a big difference between the short and long-run elasticity. In the short-run, elasticity is almost zero. In the long-run, the value of water is restricted from above by the cost of recirculation.

Another approach for measuring industrial water demand is using mathematical optimization models. This is described in detail in Kindler and Russel (1984), where the different water use technologies in a given industry are specified. Then, by varying the water constraints, a value can be imputed through cost saving within different technologies. Thus, depending on the water availability, a demand function for water can be traced from the dual values associated with the different solutions.

The value of water for residential use can be split into two main values: those connected with the quantity aspect and those dealing with the quality. The quantity issue tries to capture the value of water given a fixed quality. On the other hand, studies dealing with the quality aspects try to value the WTP for a fixed amount of water in varying qualities. The method used clearly depends on the services which are being estimated.

For the quantity aspect, when appropriate data are available, the econometric approach to estimating the demand function for water is preferred. The dependent variable is the water consumed, while the explaining variables are the price of water, income, climatic factors, house characteristics (number of bath tubs, etc.) and number of people in the household. A lack of price variation does not usually enable a time series to be estimated, but rather a cross-sectional data set. In addition, quantity is rarely measured on a per household basis, but rather on an average site basis. The price perception and simultaneity are two other main statistical problems. After these issues have been dealt with, the target is to find the price elasticity of water in order to estimate its value for residential uses.

Price elasticity exhibits a large intra-seasonal variation as well as spatial and inter-temporal (long-run vs. short-run) variation. Gibbons (1986) gives examples of price elasticity estimated for Tucson, Arizona which range from between −0.23 and −0.7 for the winter and summer respectively. In Raleigh, North Carolina winter and summer elasticity ranges between −0.3 and −1.38.

Carver and Boland's study (1980) is an important one. This study measured residential demand for water in the Washington, DC, metropolitan area using cross-section, time series and panel data. This allowed Carver and Boland to separate short-run from long-run elasticity. Elasticity estimates range between −0.05 (November to April, cross-section, short-run) to −0.7 (November to April, cross-section, and long-run). For policy purposes, this is important because demand management options will vary depending on the season.

Based their study on an extensive, cross-sectional sample of microdata from metropolitan Denver (Colorado), Jones and Morris (1984), estimated residential water demand which incorporated instrumental price variables for the average price and the variables of the two-part price specification. A strategy of instrumental estimation was employed to identify a new variable or variables correlated with price but orthogonal to the disturbance term of regression.

Rizaiza (1991) conducted a socio-economic survey in Saudi Arabia to collect the information needed for estimating a functional relationship between residential water usage and the relevant independent variables. Three different models of annual residential water usage per household (household served by tankers, those served by the public water network, and the pooled groups) were estimated by ordinary least squares. The price elasticity, along with a 95% confidence limit, were very similar to those estimated for the United States (Howe and Linaweaver, 1967; Howe, 1982).

4.3.4. *Methods based on proxy markets*

By proxy markets, we mean markets which are related to the missing market: a good correlation and a well-functioning market can provide a lot of insight.

The travel cost method (TCM)

This method is known for estimates of recreational uses of environmental resources. It is based on the cost of arriving at a specific site as a proxy for the willingness to pay for recreational benefits. In this respect, groundwater resource may indirectly provide recreational benefits, by recharging surface waters and sustaining wetlands and other recreational resources. The TCM's theoretical foundations, and how to interpret the empirical results, can be found in Freeman (1993).

The hedonic price method (HPM)

This approach applies to another class of cases for which proxy markets can provide some information on the willingness to pay and the benefits of environmental assets. Here, the surrogate market is the market for property values. In practice, there is an hedonic function which relates the property price to its attributes where water quality is one of them. If we have a relative large sample with a good variation in prices for areas that are affected and areas that are not, then the contribution of clean or reliable water from the adjacent aquifer can be monetized.

Averting cost method (ACM)
This approach is based on the household production function. It is assumed that a household produces goods using inputs. Some of them are subject to pollution. The household's response is to engage in averting behaviour. This may take several forms: buying water filters (durable goods), buying bottled water (non-durable goods) and changing daily behaviour such as boiling the water or reducing the length and frequency of showers. A theoretical foundation is found in Bartik (1988). It should be noted, however, that averting behaviour does not measure the total willingness to pay for water quality since the later is probably higher then the averting expenditures. The results of these studies should only serve as a bottom limit and should be coupled with other valuation techniques (Abdalla, 1994).

Cost of illness (COI)
Savings from possible expenses relating to illness can be used in order to estimate the benefits of groundwater pollution abatement. These costs are composed out of direct costs of medical treatment and opportunity costs such as lost earnings. Here, again, this approach only has value in setting a bottom limit since it does not represent the real willingness to pay for the discomfort.

4.3.5. *Application of methods based on proxy markets*

To our knowledge, one study has applied the TCM to assess the value of groundwater. Kulshreshtha (1994) gave an estimation of the value of groundwater in Manitoba (Canada), reached indirectly through the value of recreational activities. Using data for three regional and provincial parks, the average willingness to pay for a day-visitor recreational experience was estimated to be $4.46 in 1986 dollars. Adjusting this figure to the increase in the cost of living, one arrives at a value of $5.17/person/day. Multiplying it by the estimated number of visitor-days, the total value of the recreational experience is estimated at $26 161 per annum. The difficulty of determining the portion of recreational value attributable to groundwater is one of the major weaknesses of TCM.

Malone and Barrows (1990) and Page and Rabinowitz (1993) conducted hedonic price studies on groundwater contamination problems. The first study investigated the effect of groundwater pollution on residential property values in Portage County (Wisconsin). They used statistical methods to isolate the effects of nitrate and aldicarb pollution on property values. The results did not support the hypothesis that the higher the level of nitrate contamination, the lower the price of the residential property. Page and Rabinowitz used case studies to analyse the effect of groundwater contamination on both commercial and industrial and residential property values. The cases showed that groundwater contamination significantly influenced the value of commercial and industrial property but they also found that residential property markets behave differently to commercial property markets in response to groundwater contamination.

According to the National Research Council (1997), all these studies have had limited results in isolating the effect of polluted groundwater upon residential property values. Moreover, the data requirements for an hedonic study are large, only WTP values for small changes in groundwater quality can be accurately estimated from this approach, and the functional form of the true underlying hedonic pricing equations is unknown.

Few studies have used ACM to measure household level costs associated with groundwater pollution. Raucher (1986) developed a model to measure the benefits of groundwater protection within a probabilistic damages-avoided context. Damages avoided are calculated in a linear form with the caveat that the associated social benefits were understated to an extent inversely related to the actual degree of risk aversion. The damages-avoided approach also understated the benefits of containment and detection policies because intrinsic values related to efforts at avoiding groundwater contamination were omitted.

Abdalla (1990) estimated the economic losses from groundwater contamination in a central Pennsylvania community. The data for estimating the benefits of non-marginal reduction pollution were obtained from two sources: the household averting expenditures were collected via a mail survey, and information for calculating the upper limit benefit measurement was taken from the results of surveys of water treatment industry firms. Expenditure was estimated at \$148 900 over the 6-month contamination period or approximately \$252/household/year. The costs underestimated the lower limit measurement of welfare losses associated with groundwater contamination. The upper limit measurement of welfare losses to households from contamination was estimated at \$383/household/year.

In southeastern Pennsylvania, Abdalla *et al.* (1992) used averting expenditure to approximate the economic costs to households in a community affected by groundwater contamination. Mail questionnaires were used to elicit information about increases in household averting expenditure undertaken in response to contamination in Perkasie. Averting expenditure was estimated to range from \$61 313.29 to \$131 334.06 during the 88 weeks of contamination. Under specific assumptions, the change in averting expenditures associated with a change in environmental quality provided a conservative estimate of the true cost (or benefit) of the environmental change. This study indicated that the method was capable of yielding conceptually valid estimates of an important category of costs associated with environmental pollution.

Recently, Yadav and Wall (1998) compared the actual costs of the promotion and adoption of agricultural best management practices in the Garvin Brook Watershed (Southeastern Minnesota) area with the potential benefits of reducing groundwater nitrate concentrations. They used the avoidance costs of treating water as proxies for estimating the benefits of restoring groundwater quality. They estimated the potential annual monetary benefits of groundwater quality improvement in the project area at \$59 000 under the current situation. The cost of groundwater protection implemented was estimated to be \$57 700. Under the current level of contamination, avoidance costs would be equal to BMP programme costs in about 6 years. The best solution for this area would be to replace certain wells posing immediate health threats and at the same time to implement groundwater protection measures.

Blomqvist and Whitehead (1995) studied the value of averting seawater intrusion in the Orange Country in California using this method. Loss of the basin beyond any possible use would require the district to rely on imported water for its entire water supply. Groundwater is generally less expensive than imported water, primarily because of the development and transmission costs of the imported supplies. He estimated that the value of Orange County's groundwater over a 20-year period would be approximately \$1.39 billion, and the value of imported water would be as high as \$4.80 billion. It indicated that the present value difference of the two scenarios was approximately \$3.41 billion and this was one measure of the value of the groundwater basin, although it presumably represented a lower limit estimate of the true value. In another study by Krulce *et al.* (1997), groundwater in the Pearl Harbor aquifer was modelled as a renewable resource and as replaceable at a fixed cost by backstop resource (desalination). They adopted an efficiency simulation which assumed that the cost of extraction would rise as the head was lowered and found that from the beginning of the simulation to the time that the steady state was reached, the efficiency price increased from about four to six times the extraction cost.

Cummings and McFarland (1974) have conducted other studies on groundwater and salinity control.

4.3.6. *Methods based on direct valuation*

There are many situations in which no value measurement can be derived from observing individual choices through a proxy market. In such cases, there is no choice left but to directly ask people about their maximum willingness to pay for a possible improvement in the environmental quality of the given resource.

The most common approach is known as the contingent valuation method (CVM). The advantage of this approach is that it is the only one that allows the measurement of non-use values. However, CVM may involve several types of errors. The most important are those associated with strategic bias (respondents have a specific intention to over or under estimate their WTP), starting point bias (where the respondent is looking for clues of which one of them is the starting point) and mispecification bias (in which the respondent has a different picture of the environmental good to that intended by the researcher) (Freeman, 1993).

However, CVM studies have the advantage of directly assessing the social benefit of a public programme through the estimated WTP. For groundwater purposes, this is quite important since a large part of the services it provides are not associated with a well-functioning market, if any at all. In the next section, we will describe the major studies done in this area with the help of CVM, especially with regard to the WTP for improving the standards of drinking water.

4.3.7. *Application of methods based on direct valuation*

In the last decade, intensive research efforts have been put into estimating the value of groundwater quality. A large number of studies into the different causes of groundwater pollution are carried out using the CVM, given its ability to measure all components of economic value.

Shultz and Lindsay (1990) estimated the economic value for a hypothetical groundwater protection plan in Dover where the cause of the pollution was the leaching of chemicals and toxic wastes. A contingent valuation method experiment using the total design method was administered via a mail questionnaire. A logit technique was applied for analysing the relationship between the dichotomous and the independent variables that were collected. The mean WTP value of $129, which is associated with the truncation level at the highest bid offered ($500), can be considered the best representation of mean WTP for groundwater protection in Dover.

An assessment of groundwater subject to contamination by toxic chemicals and diesel fuel was carried out by Powell (1991) producing a mean WTP of $61.55/household/year.

Another investigation of the national benefits of cleaning groundwater contaminated by landfills was carried out by McClelland *et al.* (1992). To explore this issue, they constructed a model of intergenerational choice which assumed that the utility of the present generation depended on the utility of future generations. Econometric analysis of a national mail survey was used to correct for possible measurement error using a Box–Cox transformation. Three alternative approaches for calculating non-use values were used to provide remarkably consistent estimates of such values. The mean values for the WTP for complete groundwater cleanup (in the full sample scenario) were about $84/household/year.

Jordan and Elnagheeb (1993) used the CVM payment card for measuring the value of drinking water subject to contamination by nitrates. They found that the mean WTP was about $146/household/year for the public water systems and about $169/household/year for the private wells.

Another groundwater contingent valuation study that tested individual wells for nitrates was conducted by Poe (1998). A two-stage survey design was created in order to test individual wells and obtain values based on well test results. This study suggested that CVM research on groundwater quality and other environmental risks had adopted a paradigm that WTP values should be based on actual exposure levels. Moreover, it emphasized that values based on partial information will provide limited and biased information to decision makers.

Powell *et al.* (1994) investigated the use of contingent valuation information as a tool to persuade local government decision-makers to implement water supply protection policies. Respondents were told that a water supply protection district would be established, and all those benefiting from such a district would be asked to pay by means of an increase in their water utility bills. A major drawback of the study design was the fact that information was collected by mail questionnaire with no system for checking if respondents have read the CV information before filling out the questionnaire. A mean WTP of $61.55/household/year was revealed.

In Europe, Press and Söderqvist (1996) and Stenger and Willinger (1998) adopted the CVM for measuring the value of groundwater quality. The first study explored the economic value of groundwater resources in the Milan area (Italy). A contingent valuation method was selected for the Milan case study in order to also consider non-use values directly. An application of this method also permitted the researchers to focus specifically on the quality characteristics that were most relevant from a policy point of view in terms of pollution control options. All the assumptions made resulted in the following estimate of mean annual household WTP of about ITL645 000 ($371 in 1998 dollars).

Stenger and Willinger (1998) estimated the value that households assigned to the preservation of the quality of the Alsatian aquifer (France). They chose this technique because groundwater quality was a complex mixture of future use value, option value, bequest value and existence value. Respondents were interviewed face to face and had to respond to a yes/no question followed by an open-ended question. One problem with this study was the scarce visibility that groundwater quality had for the respondents. The observed mean WTP was equal to 617FF ($104 in 1998 dollars) per household per year. The different regressions done with the stated WTP for the open-ended method gave mean WTP estimates of between 610FF ($103 in 1998 dollars) and 709FF ($120 in 1998 dollars).

Edwards (1988) used a contingent valuation method to collect data on option prices to protect a 'sole source' aquifer from uncertain future nitrate contamination. Different results were found: first, the sensitivity of option prices to a change in the probability of supply indicates that in at least this case, the benefits of an aquifer management project should not be calculated only from certain changes in the availability of the resource. A second (and surprising) result was the small size of the option value. The small size of the option value in this study suggests that the benefits of aquifer management can be almost completely measured in terms of an increase in the expected value of benefits. A third interesting result is the strong influence of bequest motives on total willingness to pay. Finally, these results further illustrate the fact that a benefit–cost analysis of groundwater problems is inherently site specific. Each of these effects on option price should be evaluated separately for homogeneous units.

Another study by Tsur and Zemel (1995) studied the effects of irreversible uncertain events on the exploitation of groundwater resources. Irreversible events are situations in which the resource can no longer be used. The uncertainty in this model was partly exogenous, so the event might occur regardless of whether the pollution stock increases, decreases, or remains constant. The analysis is carried out by establishing a relationship between the equilibrium states and the roots of simple functions of the state variable that depend on the structural relationships and parameters.

Quantitative and detail studies about the ecological value of groundwater are still rather scarce. Evaluation of the ecological impacts is highly dependent on the social perception of ecological values in the corresponding region. Troyak (1996) has studied the total ecological and economic value of groundwater in the town of Caledon, Canada. He gave a quantitative valuation only for the use value of groundwater (between \$9.6 and \$33 million in 1995), affirming that it is much less tangible, more subjective, and more difficult to approximate dollar value for the non-use values of groundwater. From a theoretical point of view and with the knowledge accrued from the different case studies, CVM could be the only technique capable of valuing these kind of groundwater services.

According to the different CVM studies on groundwater, it can be claimed that the method presents some drawbacks because of the nature of the resource, which has scarce visibility for the respondents and so the values obtained could be based on partial information that in turn could provide limited and biased policy direction to decision makers.

4.3.8. *Other methods and their application*

There are other estimation techniques that do not fall in any specific category. We will briefly mention them here.

Dynamic optimization

Here the problem is formulated in a dynamic form through the equation of motion of water in the aquifer. The importance of this is its ability to capture the scarcity value of the stock by finding the shadow price of the aquifer (Negri, 1990). This value is associated with the aquifer's water table. Here, a dynamic analysis taking into account the flow equation of the aquifer is in order:

$$S_{t+1} = S_t + R_t - W_t \tag{5}$$

where S_t = stock level at time t, R_t = recharge at time t, W_t = withdrawal at time t.

We also know that the cost of abstraction is a function of the pumping technology, energy cost and the water table. Assuming the first two are fixed, then the cost (C) is inversely related to the water table.

$$\partial C / \partial S < 0 \tag{6}$$

The stock value is a measurement of the cost of lowering the water table now and hence, increasing the cost of abstraction for all future periods. The current water use will be efficient and in a steady state only if two conditions hold:

(a) water withdrawal is equal to water recharge;
(b) the marginal benefit of water is equal to the discounted extra costs associated with lowering of the water table due to the use of that marginal unit of water. This discounted cost is the stock value.

The first author to deal with this subject was Burt (1970). He shows that under perfect competition, the individual pumping water out of the aquifer will not take that value into account provided the individual's demand is small relative to the total demand from the groundwater. Therefore, there is a spillover effect on all the other users, which implies a lower water table than efficient water use would dictate. How large this effect is is a function of the parameters of the problem. These include extraction cost, benefits from extraction, storage facility and the interest rate.

Cummings and McFarland (1974) developed a model that links the water table to salinity and derives the stock value of this. Gisser and Sanchez (1980) developed this model and looked at the difference between unrestricted and the optimal water use. Interestingly, they found out that as long as the aquifer is large enough relative to the overall demand, then there is not much difference between the two cases – up to 5%, depending on the different assumptions (Gissar, 1983). Alternatively, Feinerman and Knapp (1983) show that Gissar's analysis is restricted to a high discount rate and low water demand. Relaxing these assumptions would dramatically change the stock value (which in their study is called 'the benefit from management'). Thus, as mentioned before, it is a question of the relevant parameters.

There are numerous other studies that have tried to evaluate the stock value. Negri (1990) presents a hypothetical example, Becker and Easter (1992) found that about 30% of the water externalities in the Great Lakes could be associated with the stock value. Similar results where found in other studies, such as that by Mueller (1983) for the Ogallala aquifer.

In addition to the stock value, it would also prove to be useful to evaluate the aquifer's buffer value. Tsur and Graham-Tomasi (1991) used dynamic programming methods to estimate the buffer value of groundwater. They highlighted the potential for uncertainty in surface water availability to affect groundwater extraction over time, explicating the buffer role of groundwater. This influence depends on the size of the aquifer stock, its extraction cost, and uncertainty. A series of options or scenarios to reduce aquifer use in the long-term and by limiting use to periods of extended drought were examined. They found that this value can be significant. Where small variability in surface water and relatively large aquifer stock exist, the buffer value accounts for 5% of the value of groundwater. In cases with high variability in the supply of surface water, a smaller aquifer and higher unit pumping cost, the buffer value accounts for 84% of the total value of the groundwater stock, and if this value were ignored, groundwater would be severely undervalued. The presence of a positive buffer value implies that groundwater is more precious in uncertain environments than in stable ones, and the difference represents the buffer value.

According to the National Research Council (1997), two case studies dealing with the buffer value of groundwater were carried out in Albuquerque, New Mexico, and the Arvin-Edison Water Storage District in southern California. The first case examines a series of options or scenarios for the city to reduce aquifer use to a long-term sustainable level by limiting use to periods of extended drought. Alternative strategies implied costs to present users but with potential long-term benefits. To weight the benefits and costs of alternative actions required the measurement over time of the economic value of an array of services within the range of options available to the city. In the second example, in California, less-than-average precipitation occurs with a frequency of about four years out of seven. To the extent that precipitation shortfalls are reflected in reductions in deliveries of surface water, groundwater buffering values will be realized in each year that precipitation is less than average. The magnitude of the value will depend upon the degree to which surface water deliveries are deficient. Because of water stored in the aquifer not being available, a rough calculation suggests that in 1991, more than 26 000 acres would have been left fallow. Assuming typical cropping patterns and typical prices, the gross value of production on this acreage would have exceeded $38 million. The returns of fixed and operating costs to the growers' net were almost $6 million.

Benefit transfer

This approach employs results from primary studies as secondary data sources and uses regression techniques to estimate the specific site benefit. We will not define this approach as an estimation approach, but rather a statistical way to verify the validity of the estimate in the new site. The major benefit of this approach is that it saves the resources needed in order to engage in a new original study. However, measurement errors in the original studies may be compounded when using old estimates for a new site.

Crutchfield *et al.* (1995) illustrated how this type of method could be carried out in the groundwater protection context. The authors examined the available groundwater valuation literature to identify benefit estimates for possible application in their research, and they found three studies. For two of the three studies considered, the estimated value based on transfer of the valuation function was about the same as the value based on transfer of the mean willingness to pay. Aggregate willingness to pay for groundwater protection was estimated at $197–730 million/year.

Troyak (1996) estimated the value for the popular recreational sport of cold-water fishing in Caledon (Canada) because the consistently cool temperatures of groundwater are necessary in creating a suitable habitat for this type of activity. He applied a benefit transfer method to determine the groundwater's value to cold-water fishing. Using a prior estimate, this was approximately $1.3 million in 1995 or approximately $34 per capita. According to the author, this was a conservative estimate since it considered only one area and did not consider other indirect contributions to recreational activities by the groundwater.

Table 4.3 summarizes the different papers described in Section 3 and is divided according to the type of estimation and what value has been estimated.

4.4. Final remarks

This chapter provides a general framework for analysing the social value of groundwater resources. For policy purposes, what is more important are the impacts derived from changing the flow and stock of services provided by an aquifer. The most important point is that the groundwater services could be enjoyed by more than one party and that preferences in regard to these services often conflict.

We have listed all the services that we believe should be accounted for when a groundwater management policy is considered to be in order, or when a change in the policy is being considered. Some of the services are pretty easy to assess since they rely on market transactions; others do not rely on market forces but still have a value associated with them. Furthermore, there may be situations in which the only possibility of measurement is to ask people to state their maximum WTP for a possible improvement in the environmental quality of the given resources.

We can conclude that for all we know there is an unbalance between the different groundwater valuation areas. In our opinion, an additional research effort should be put into addressing those groundwater functions that have only recently been recognized as being valuable (for example all the environmental groundwater services). On the other hand, in semi-arid regions, where the average is sufficient but the variance is an issue, the increasing importance of an aquifer as a buffer should be recognized and given its correct priority. This point is likely to become even more important if and when global warming takes place.

A derived result is that more comprehensive research efforts should take place within the European Union. As can be seen from the reference list and the papers surveyed, the majority of them are not from the European continent but from the US and semi-arid regions of the world. As water is becoming increasingly more scarce (both quantitatively and qualitatively), the economic factors determining the most efficient allocation and extraction will also become increasingly more important.

Table 4.3. Groundwater valuation studies

Study	Goods being valued	Description of valuation procedure
Demand function for agricultural water use (quantity aspects)		
Letey (1991), Becker (1995), and Howeitt and Vaux (1995)	Value of irrigated water	Land allocation among crops
Becker (1995)	Value of the Mountain Aquifer (Israel)	Given the total amount of water in the aquifer, different allocation mechanisms can be determined
Boggess *et al.* (1993)	Elasticity of water	Programming model
Moore and Dinar (1992)	Water demand through estimate of later allocation among crops	Quantity-shadow price relationship
Dinar and Letey (1994), and Boggess *et al.* (1993)	Impact of changing input level on the output	Water-crop production function
Crop response model for agricultural water use (quality aspect)		
Letey (1991)	Salinity abatement	Mathematical programming in order to find a production function reflecting the change in yields due to reduced salinity
Demand function for industrial water use		
Babin *et al.* (1982), and Renzetti (1992)	Industrial water use in Canada	Cobb–Douglas production function
Kindler and Russel (1984)	Industrial water demand	Mathematical optimization models
Demand function for residential water use		
Gibbons (1986), Carver and Boland (1980), Jones and Morris (1984), and Rizaiza (1991)	Estimation of residential water demand	Econometric approaches

Study	Goods being valued	Estimated willingness to pay	Description of valuation procedure
CVM case study of groundwater protection quality			
Randall *et al.* (1983)	Protection of groundwater subject to pesticides and nitrates	Rural: \$43–46/hh*/year Urban: \$34–69/hh/year	Open questions in CVM
Powell (1991)	Groundwater subject to contamination by toxic chemicals and diesel fuel	All data: \$61.55/hh/year Respondents with a history of contamination: \$81.66/hh/year Respondents with no contamination: \$55.79/hh/year	Method of computation not specified. WTP for private well users exceeds WTP for public water supply users by \$14.04
Shultz and Lindsay (1990)	Groundwater, type of contaminant not specified	Mean WTP: \$129/hh/year	Computed from logit model
Jordan and Elnagheeb (1993)	Drinking water subject to contamination by nitrates	Public water systems: \$146/hh/year. Private wells: \$169/hh/year	Averages computed at midpoints from CVM payment card

Table 4.3. (*Continued.*)

Study	*Goods being valued*	*Estimated willingness to pay*	*Description of valuation procedure*
Sun *et al.* (1992)	Groundwater subject to contamination by agricultural fertilizers, nitrates and pesticides	Mean WTP: \$641/hh/year, ranges from \$165–1452/hh/year	Computed from logit model
Stenger and Willinger (1998)	Groundwater subject to contamination by intensive use of fertilizers in agriculture	Mean WTP: 1200 FF hh/year	Computed from logit model (dichotomous choice method)
Press and Söderqvist (1996)	Groundwater subject to contamination by pesticides	Mean WTP: Lit 645 000/hh/year	Computed from logit model with truncation
CVM case study of option value			
Edwards (1988)	Groundwater subject to contamination by nitrates and pesticides	\$286–1130/hh/year	Derived from figure 2 published in the journal article by Edwards
CVM case study of non-use value			
McClelland *et al.* (1992)	Groundwater, type of contaminant not specified	Complete sample: \$84/hh/year	Predictions from Box–Cox model
ACM on groundwater protection quality			
Abdalla *et al.* (1992)	Groundwater subject to contamination by a volatile synthetic organic chemical (TCE)	\$61 313–131 334 during 88 week of contamination period	Estimated costs due to TCE contamination from Dec. '87 to Sept. '89
Abdalla (1990)	Groundwater subject to perchloroethylene (PCE)	\$252/hh/year	Cost estimated empirically for a community served by a public water system
Raucher (1986)	Groundwater affected by land-based waste disposal in three different case studies (two in Florida and one in new Hampshire)	\$0.64 million, \$127.4 million and from \$6.2 million to \$148.8 million	Estimated cost per cancer avoided with different time horizons
Hedonic price method on groundwater quality			
Malone and Barrows (1990)	Groundwater contamination problem	The results do not support the hypothesis that the higher the level of nitrate contamination, the lower the price of the residential property. However, they do not accept the hypothesis that nitrate levels have no economic effect in the housing market because the market may react to nitrate in several ways	Regression in the effect of nitrate levels on residential property values in Portage County, Wisconsin

Table 4.3. (*Continued.*)

Study	*Goods being valued*	*Estimated willingness to pay*	*Description of valuation procedure*
Page and Rabinowitz (1993)	Groundwater contamination problem	They found no measurable effect on property values in the residential property case studies, in contrast to the commercial and industrial case studies	They used case studies to analyze the effect of groundwater contamination on commercial, industrial and residential property values
Benefit transfer analysis of groundwater quality			
Crutchfield *et al.* (1995)	Groundwater protection	Aggregate WTP for groundwater protection was estimated at $197–730 million per year. The household values were multiplied by the number of rural households in each country considered	Using 4 policy sites and computed mean values of the independent variables on a county-by-county bases
Dynamic programming methods for buffer value			
Tsur and Graham-Tomasi (1991)	Evaluation of the buffer value to wheat growers of the fossil aquifer underlying the Northern Negev Region	The buffer value consists of between 5% and 84% (small, high variability) of the value of groundwater	The definition of buffer value is based on value functions for an intertemporal–optimization problem

* hh = household.

Notes

1. Aquifer and groundwater resources are used interchangeably in this chapter.
2. We need to take into account that we are dealing here only with the benefit side of the programme and not the direct cost. However, part of the benefit consists of avoiding implicit costs such as sea water intrusion, etc.

References

Abdalla, C.W., 1990, Measuring economic losses from groundwater contamination: an investigation of household avoidance costs, Water Resources Bulletin, 26, 451–462.

Abdalla, C.W., 1994, Groundwater values from avoidance cost studies: implications for policy and future research, American Journal of Agricultural Economics, 76, 1062–1067.

Abdalla, C.W., Roach, B.A., Epp, D.J., 1992, Valuing environmental quality changes using averting expenditures: an application to groundwater contamination, Land Economics, 68, pp. 163–9.

Al-Qunaibet, M.H., Johnston, R.S., 1985, Municipal demand for water in Kuwait: Methodological issues and empirical results, Water Resources Research, 22, 401–409.

Amigues, J.P., Gaudet, G., Moreaux, M., 1997, Rational pricing of groundwater with a growing population, Environmental Modeling and Assessment, 2, 323–331.

Babin, F., Willis C., Allen, P., 1982, Estimation of substitution possibilities between water and other production inputs, American Journal of Agricultural Economics, 68, 647–656.

Bartik, T.J., 1988, Evaluating the benefits of non-marginal reductions in pollution using information on defensive expenditures, Journal of Environmental Economics and Management, 15(1), 111–127.

Becker, N., Easter, K.W., 1992, Dynamic supply from a common property resources: Water diversions from the Great Lakes, Natural Resource Modeling, 6, 343–371.

Becker, N., 1995, The value of institutional change in moving from central planning to a market system: Implications from the Israeli water sector, Agricultural Economics, 12, 11–21.

Becker, N., Zeitouni, N., Schecter, M., 1996, Reallocating water resources in the Middle East through market mechanisms, International Journal of Water Resources Development, 12(1), 17–32.

Beckerman, W., 1995, Small Is Stupid: Blowing the Whistle on the Greens, London: Duckworth.

Blomqvist, G.C., Whitehead, J.C., 1995, Existence value, contingent valuation, and natural resources damages assessment, Growth and Change, 26(4), 573–589.

Boggess, W., Lacewell, R., Zilberman, D., 1993, Economics of water use in agriculture, in: Carlson, G.A., Zilberman, D., Miranowski, J.A., eds., Agricultural and Environmental Resource Economics, New York: Oxford University Press.

Booker, J.F., Young, R.A., Modeling intrastate and interstate markets for Colorado River water resources, Journal of Environmental Economics and Management, 26, 66–87.

Boyle, K.J., Poe, G.L., Bergstrom, J.C., 1994, What do we know about groundwater values? Preliminary implications from a meta-analysis of contingent-valuation studies, American Journal of Agricultural Economics, 76, 1055–1061.

Burt, O.R., 1970, Groundwater storage control under institutional restrictions, Water Resources Research, 6(6), 1540–1548.

Carver, P.H., Boland, J.J., 1980, Short run and long run effects of price on municipal water use, Water Resources Research, 16, 609–616.

Colby, B., 1989, Estimating the value of water in alternative uses, Natural Resources Journal, 29, 511–527.

Crocker, T.D., Forster, B.A, Shogren, J.F., 1991, Valuing potential groundwater protection benefits, Water Resources Research, 27, 1–6.

Crutchfield, S.R., Feather, P.M., Hellestein, D.R., 1995, The Benefits of Protecting Rural Water Quality. An Empirical Analysis, Economic Research Service, Agricultural Economic Report No. 701, United States Department of Agriculture.

Cummings, R.G, McFarland, J.W., 1974, Groundwater management and salinity control, Journal of Environmental Economics and Management, 10, 909–915.

Daly, H.E., On Wilfred Beckerman's critique of sustainable development, Environmental Values, 4, 49 55.

Dinar, A., Letey, J., 1994, Evaluation of management and policy issues related to irrigation and agricultural crops, in: Tanji, K.K., Yaron, D., eds., Advanced Series in Agricultural Sciences 22, Berlin/Heidelberg: Springer.

Edwards, S.F., 1988, Option prices for groundwater protection, Journal of Environmental Economics and Management, 15, 475–87.

Feinerman, E., Knapp, K.C., 1983, Benefits from groundwater management: magnitude, sensitivity, and distribution, American Journal of Agricultural Economics, 65(4), 703–710.

Freeman, A.M. III, 1985, Supply uncertainty, option price, and option value, Land Economics, 61, 176–81.

Freeman, A.M. III, 1993, The Measurement of Environmental and Resource Values: Theory and Methods, Washington, DC: Resources for the Future.

Gibbons, D., 1986, The Economic Value of Water, Washington, DC: Resources for the Future.

Gisser, M., Sanchez, D.A., 1980, Competition versus optimal control in groundwater pumping, Water Resources Research, 16, 638–642.

Gisser, M., 1983, Groundwater: focusing on the real issue, Journal of Political Economy, 91, 1001–1027.

Goldsmith, E., Allen, R., Allaby, M., Davoll, J., Lawrence, S., 1972, Blueprint for survival, The Ecologist, 2, 1–40.

Haang, W.Y., Uri, N.D., 1990, An analytical framework for assessing the benefits and costs of policies related to protecting groundwater quality, Environment and Planning, 22, 1469–1486.

Howe, C.W., Linaweaver, F.P., 1967, The impact of price on residential water demand and its relation to system design and price structure, Water Resources Research, 3, 13–32.

Howe, C.W., 1982, The impact of price on residential water demand: some new insights. Water Resources Research, 18(4), 713–716.

Howeitt, R.E., Vaux, H., 1995, Competing demands for California's scarce water, in: Dinar, A, Loehman, E.T., eds., Water quantity/quality management and conflict resolution: institutions, processes, and economic analyses. Kluwer Academic Publishers, 271–287.

Johnson, S.L., Adams, R.M., Perry, G.M., 1991, The on-farm costs of reducing groundwater pollution, American Journal of Agricultural Economics, 73(4), 1063–1073.

Jones, C.V., Morris, J.R., 1984, Instrumental price estimates and residential water demand, Water Resources Research, 20, 197–202.

Jordan J.L., Elnagheeb, A.H., 1993, Willingness to pay for improvement in drinking water quality, Water Resources Research, 29, 237–45.

Kim, C.S., Moore, M.R., Hanchar, J.J., Nieswiodomy, M., 1989, A dynamic model of adaptation to resource depletion: theory and an application to groundwater mining, Journal of Environmental Economics and Management, 17, 66–82.

Kindler, I., Russel C.S., eds., 1984, Modeling Water Demands, New York: Academic Press.

Knapp, K.C., Olson, L.J., 1995, The economics of conjunctive groundwater management with stochastic surface supplies, Journal of Environmental Economics and Management, 28, 340–356.

Krulce, D.L., Rumasset, J.A., Wilson, T., 1997, Optimal management of a renewable and replaceable resource: the case of coastal groundwater, American Journal of Agricultuaral Economics, 79, 1218–1228.

Kulshreshtha, S.N., 1994, Economic Value of Groundwater in the Assiniboine Delta Aquifer in Manitoba. Social Science Series no. 29. Environmental Conservation Service. Ottawa, Ontario: Minister of Supply and Services.

McCarl, B.A., 1982, Cropping activities in agricultural sectors: a methodological proposal, American Journal of Agricultural Economics, 64(4), 768–772.

Moore, M.R., Dinar, A., 1995, Water and land as quantity-rationed inputs in California agricultural empirical tests and water policy implications, Land Economics, 71(4), 445–461.

Letey, J., 1991, Crop-water production functions and the problems of drainage and salinity, in: Dinar,A., Zilberman, D., eds., The Economics and Management of Water and Drainage in Agriculture. Kluwer Academic Publishers.

Lichtenberg, E., Zilberman, D., Bogen, K.T., 1989, Regulating environmental health risks under uncertainty: groundwater contamination in California, Journal of Environmental Economics and Management, 17, 22–34.

Malone, P., Barrows, R., 1990, Groundwater pollution effects on residential property values, Portage County, Wisconsin, Journal of Soil and Water Conservation, 45, 346–48.

McClelland, G.H., Schulze, W.D., Lazo, J.K., Waldman, D.M., Doyle, J.K. Elliott, S.R., Irwin, J.R., 1992, Methods for measuring non-use values: a contingent valuation study of groundwater cleanup, Mimeo, Center for Economic Analysis, University of Colorado. USEPA no. CR-815183.

Mitchell, R.C., Carson, R.T., 1989, Using Survey to Value Public Goods: The Contingent Valuation Method, Washington, DC: Resources for the Future.

Moore, M.R., Dinar, A., 1995, Water and land as quantity-rationed inputs in California agriculture empirical tests and water policy implications, Land Economics, 71(4), 445–461.

Mueller, M.J., 1983, Optimal groundwater depletion with interdependent nonrenewable and renewable resources, Water Resources Research, 19, 1091–1098.

National Research Council, 1997, Valuing Ground Water: Economic Concepts and Approaches, National Academy Press, Washington, D.C.

Negri, D.H., 1990, 'Stragedy' of the commons, Natural Resource Modeling, 4, 521–537.

Norton, B.G., 1995, Evaluating ecosystem states: two competing paradigms, Ecological Economics, 14, 113–127.

Norton, G.B., Toman, M.A., 1997, Sustainability: ecological and economic perspectives, Land Economics, 73, 553–568.
Page, G.W., Rabinowitz, H., 1993, Groundwater contamination: its effects on property values and cities, Journal of the American Planning Association, 59, 473–81.
Poe, G.L., 1998, Valuation of groundwater quality: contingent values, public policy needs, and damage functions, ARME Working Paper 97–22.
Powell, J.R., 1991, The value of groundwater protection: willingness to pay information and their utilization by local government decision makers. Ph.D. Dissertation, Cornell University.
Powell, J.R., Allee, D.J., McClintock, C., 1994, Groundwater protection benefits and local community planning: impact of contingent valuation information, American Journal of Agricultural Economics, 76, 1068–1075.
Press, J., Soderqvist, T., 1996, On estimating the benefits of groundwater protection: a contingent valuation study in Milan, FEEM Working Paper 53.96.
Randall, A., 1994, A difficulty with the travel cost method, Land Economics, 70, 88–96.
Randall, A., Hoehn, J.P., Brookshire, D.S., 1983, Contingent valuation surveys for evaluating environmental assets, Natural Resources Journal, 23(3), 635–648.
Raucher, R.L., 1986, The benefits and costs of policies related to groundwater contamination, Land Economics, 62, 33–45.
Renzetti, S., 1992, Estimating the structure of industrial water demands: the case of Canadian manufacturing, Land Economics, 68, 396–404.
Rizaiza, O.S.A., 1991, Residential water usage: a case study of the major cities in the western region of Saudi Arabia, Water Resources Research, 28, 667–671.
Segerson, K., 1994, The benefits of groundwater protection: discussion, American Agricultural Economics, 76, 1076–1078.
Shultz, S.D., Lindsay, B.E., 1990, The willingness to pay for groundwater protection, Water Resources Research, 26, 1869–1875.
Stenger, A., Willinger, M., 1998, Preservation value for groundwater quality in a large aquifer: a contingent-valuation study of the Alsatian aquifer, Journal of Environmental Management, 53, 177–193.
Sun, H., Bergstrom, J.C., Dorfman, J.H., 1992, Southern Journal of Agricultural Economics, 24(2), 63–71.
Sunding, D., 1995, Flexible technology and the cost of improving groundwater quality, Natural Resource Modeling, 9(2), 177–192.
Troyak, M., 1996, An Assessment of the Ecological and Economic Value of Groundwater: Town of Caledon Case Study. Canada Center for Inland Waters, Ontario Region.
Tsur, Y., Zemel, A., 1995, Uncertainty and irreversibility in groundwater resource management, Journal of Environmental Economics and Management, 29, 149–161.
Tsur, Y., Graham-Tomasi, T., 1991, The buffer value of groundwater with stochastic surface water supplies, Journal of Environmental Economics and Management, 21, 201–224.
United States Environmental Protection Agency, 1995, A Framework for Measuring the Economic Benefits of Ground Water. Office of Water, EPA 230-B-95-003.
Yadav, S.N., 1997, Dynamic optimization of nitrogen use when groundwater contamination is internalized at the standard in the long run, American Journal of Agricultural Economics, 79, 931–945.
Yadav, S.N., Wall, D.B., 1998, Benefit–cost analysis of best management practices implemented to control nitrate contamination of groundwater, Water Resources Research, 34, 497–504.
Young, R.A., Gray, S.L., 1985, Input–output models, economic surplus, and the evaluation of state or regional water plans, Water Resources Research, 21, 1819–1823.
Zekter, I.S., Belousova, A.P., Dudov, V.Yu., 1995, Regional Assessment and Mapping of Groundwater Vulnerability to Contamination, Environmental Geology, 25, 225–231.

CHAPTER 5

Misconceptions in Aquifer Over-exploitation: Implications for Water Policy in Southern Europe

Nuria Hernández-Mora, Ramón Llamas and Luis Martínez Cortina

5.1. Introduction

Groundwater is a key component of any sustainable water management policy, particularly in arid and semi-arid countries where water resources are scarce. Dosi and Tonin (Chapter 2, this volume, Tables 2.8 and 2.9) provide an overview of the importance of groundwater use by sector in southern European countries. While national averages do not account for the strong regional differences in groundwater use, it is apparent that groundwater plays a key role as an integral part of the national water management strategies in many of these countries.

Groundwater development has been particularly strong since the 1950s. Advancements in the science of hydrogeology, in well drilling technology, and the invention of the turbine pump, have all facilitated spectacular development. In Spain, for instance, it is estimated that groundwater abstraction has increased approximately five-fold since the 1950s (MOPTMA, 1995), to 5500 Mm^3 today. The growth in groundwater use has been primarily the result of initiatives taken by thousands of individual users, industries and small municipalities. The public sector has rarely participated in the planning, administration or control of these developments.

This remarkable growth has brought benefits, but in some cases has also resulted in undesired consequences for the environment, for third parties, and for the long-term sustainability of the resource itself. These negative consequences are what have come to be known as situations of "over-exploitation". Dealing with the undesired consequences of intensive groundwater use has become a necessary goal in order to ensure the long-term sustainability of the resource.

This chapter looks at the concept of sustainable use of groundwater resources and the confused concept of aquifer over-exploitation. We start by presenting the inherent difficulties of estimating renewable resources in an aquifer given data uncertainties, changing social preferences, and the impacts of human

C. Dosi (ed.), Agricultural Use of Groundwater, 107–126.

activities and different management strategies on renewable resources. We suggest that an aquifer is overexploited when the balance between the benefits and costs of a certain level of groundwater abstraction is negative considering long-term values. We then review criteria that can serve to detect the susceptibility of an aquifer to adverse effects (hydrological, ecological, geotechnical, economic). We end by suggesting some key aspects that should become an integral part of any rational groundwater management policy.

5.2. Evaluating available renewable resources

A fundamental issue that needs to be addressed prior to the design of any water management programme is the estimation of the available resources (surface and groundwater). Once this basic information is available, priorities can be set and decisions made to accommodate different uses and requirements.

It is generally accepted that water management should aim to be sustainable. However, finding a common practical definition of sustainability is not an easy task. For instance, what time frame should we consider when talking about sustainability: 50, 100, 500 years? In general, sustainable use of water resources is understood in purely hydrological terms, so that available resources in a catchment are defined as renewable resources.

In the case of groundwater in arid and semi-arid countries, defining renewable resources is particularly challenging given the difficulty of estimating recharge values and calculating total abstraction. In addition, human activities have a positive or negative impact on recharge values and management strategies can serve to maximize available resources. It is therefore important to constantly update working estimates and adapt management plans accordingly.

5.2.1. *Uncertainty in estimating available resources*

Uncertainty in estimating available resources is a result of the scarcity and poor quality of hydrogeological data and the strong non-linearities in the hydrologic cycle of arid and semi-arid countries. Scarcity of good quality, long-term climatic and hydrologic data hampers the ability of decision makers to make reliable estimates of renewable resources. In the case of groundwater, good quality data is essential for determining renewable resources and aquifer responses to different levels of abstraction.

It is significant that, in spite of its importance, groundwater continues to be a largely misunderstood and often neglected resource. As Table 5.1 suggests, even today, some southern European countries lack a homogeneous nationwide monitoring system for groundwater quality and quantity. When monitoring does occur, data are often insufficient or unreliable and record lengths usually short. Finally, public access to existing data is not always easy and can sometimes be expensive. This is the case in Spain, for instance, where fees are charged

Table 5.1. Groundwater monitoring systems in southern EU

Country	Groundwater quality	Groundwater quantity
France	Coverage: National Origin: 1902 Average record length: 8 years No national quality control Data available free of charge	No national network Various regional networks No warranty of data quality National network planned for future
Greece	No national network Some periodic monitoring National network planned	No national network Sporadic monitoring within project frameworks Database under development
Italy	No national network Decentralized monitoring Two primary sets of networks purposes: drinking water (since 1988) environmental quality control	No national network Decentralized monitoring Some regional coordination
Portugal	No national network Only three systems monitored Origin: 1977 Average record length: 18 years No periodic reporting	Coverage: National Origin: 1970 Average record length: 15 years Monthly hydrological reporting
Spain	Coverage: National Origin: 1967 Average record length: 10 years Data quality control Data accessible without restrictions, fees apply Some reporting	Coverage: National Origin: 1967 Average record length: 10 years Data accessible without restrictions, fees apply Some reporting

Source: Original table using data from European Environmental Agency (1996).

for the use of public hydrogeological data. As will be later discussed, lack of adequate data continues to be one of the primary impediments for sustainable management.

Other uncertainties are intrinsic to the climatic variability that is particularly acute in arid and semi-arid countries. This variability results in dramatic seasonal and year-to-year fluctuations in rainfall, groundwater recharge, surface runoff, or stream flow. Future fluctuations are difficult to predict with certainty given our limited understanding of climate. In this context, making management decisions based on estimated mean annual groundwater recharge or stream flow might be risky, particularly in small basins and during dry long spells. Recharge estimates for management purposes should take these limitations into consideration.

Finally, when thinking about sustainability, it is important to keep in mind that social priorities and goals are increasingly setting the agenda for natural

Table 5.2. Anthropogenic impacts on a catchment's hydrological cycle

Impact	Effects	Availabile resources
Wetland disappearance	Decreased evaporation and evapotranspiration	Increased for other uses such as irrigation
Changes in river systems	Gaining to losing Losing to gaining	Increase Decrease
Infiltration from irrigation	Added recharge	Increase
Increase in impervious surface (urbanization, roads)	Decreased recharge* Decreased base flow in urban streams Increased surface run-off	Decrease Variable Variable
Changes in land cover	Changes in recharge	Variable
Pollution	Non-point source pollution from agricultural returns Point-source pollution from industries, sewage treatment plants, landfills, etc.	Decreased availability for certain uses

* In some cases recharge may increase because of leakages from drinking water and sewage distribution networks.
Source: Acreman *et al.* (2000).

resource management decisions. However, these priorities differ from country to country and evolve over time. A clear example of this evolution is the growing appreciation of the environmental value of water. As we will later see, draw-downs in the water table that result from groundwater abstraction can have the effect of increasing the amount of resources available for other uses. While this was seen in the past as a net benefit, the possible negative impacts that draw-downs of the water table can have on associated aquatic ecosystems has caused a reconsideration of this effect as a net benefit.

5.2.2. *Impacts of human-induced land use changes on available resources*

Human activities can have a significant impact on available resources, so that continuous monitoring and update of working estimates are necessary. A realistic goal would be to require 5-year updates and revisions of available resources and management plans, or in agreement with what is required by the European Union Water Framework Directive. Table 5.2 suggests some examples of such possible impacts and the effects they might have.

5.2.3. *Integrated management of surface and groundwater resources*

Conjunctive use of surface and groundwater resources is an efficient way of maximizing resource utilization. The concept of conjunctive use is, in principle, easy to understand. During humid periods, excess surface water is used to artificially recharge aquifers. In times of drought, groundwater is used. However, what usually takes place is an alternating use, so that surface water is the primary water source while groundwater is only used in times of drought. This approach is what has become commonplace in Spain, for instance. While this approach is acceptable, it does not maximize resource use. Carrying out active groundwater management – withdrawals in excess of recharge in the dry season or during dry sequences and aquifer recharge during the wet season or periods – should be considered, particularly in arid and semi-arid Mediterranean countries. It is the most efficient way to improve water supply guarantee, it is usually less costly than other alternatives such as new surface water infrastructures, and it is more environmentally friendly.

However, the practical implementation of such an approach can present some challenges. The technical aspects of aquifer capacity and recharge possibilities must be adequately studied. But the primary challenges are legal, economic and political. Some significant issues are:

- Who should authorize surface water used for recharge?
- Who is responsible for the costs involved in artificial recharge operations?
- Who has the right to use the recharged water?
- What organization will have management responsibilities?

These issues need to be solved before implementing a conjunctive use strategy.

5.3. The complex concept of aquifer over-exploitation

A key component in any definition of sustainable water use is a better understanding of the concept of aquifer over-exploitation. Terms relating to over-exploitation used in the literature include safe yield, sustained yield, overdraft, exploitation of fossil groundwater, and optimal yield (Adams and MacDonald, 1995; Fetter, 1994). All these terms have in common the idea of avoiding undesirable effects as a result of intensive groundwater development. However, this undesirability is largely a matter of social perceptions of the issue, perceptions which are more related to the cultural, regulatory and economic context than to strict hydrogeological data.

Many authors consider that, strictly speaking, over-exploitation occurs when abstraction is greater than or close to average recharge if long-term mean values are taken into account. However, both abstraction and recharge are difficult terms to define, particularly in arid and semi-arid countries.

A possible definition is that aquifer over-exploitation occurs when the economic, social and environmental costs that derive from a certain level of water abstraction are greater than its benefits (Llamas *et al.*, 1992). Given the multi-faceted character of water, this comparative analysis should include hydrologic, ecological, socio-economic and institutional variables. While some of these variables may be difficult to measure and compare, they must be explicitly included in the analysis so they can inform decision-making processes. Below we present seven possible criteria that can be used to evaluate the susceptibility of an aquifer to be stressed, or over-exploited. Criteria 2, 3 and 4 were proposed by Adams and MacDonald (1995). Llamas (1998) proposed criteria 5 and 6. In this paper, we will also consider the benefits of water use and the strength or weakness of institutions responsible for dealing with situations of over-exploitation.

5.3.1. *Benefits of groundwater use*

As stressed by Becker and Tonin (Chapter 4, this volume), any evaluation of available resources and the concept of over-exploitation must necessarily take into account the numerous socio-economic and even ecological benefits that can be derived from groundwater use and resource conservation. Socio-economic benefits range from water supply and sanitation to economic development as a result of agricultural growth in a particular region. With respect to potential ecological benefits, the conjunctive use of surface and groundwater resources can eliminate the need for new large infrastructure works that would seriously damage the natural hydraulic regime of a river or stream.

Water supply

Groundwater is a key source of drinking water, particularly in rural areas and in island environments. In Spain, for example, medium and small municipalities (of less than 20 000 inhabitants) obtain 70% of their water supply from groundwater sources (MIMAM, 2000). In some coastal areas and islands, the dependence on groundwater as a source of drinking water is even higher.

Irrigation

In many arid and semi-arid countries, the main groundwater use is for agriculture. Although few studies have looked at the role that groundwater plays in irrigation, those that do exist point to a higher productivity of irrigated agriculture using groundwater than that using surface water. A recent study done for Andalucía, in southern Spain (Corominas, 1999),[1] shows that, per volume of water used, irrigated agriculture using groundwater is economically over three times more productive and generates almost twice the employment than agriculture using surface water. The top part of Table 5.3 shows the results of this study. It is important to note that these results are based on the average water volumes applied to each irrigated agricultural unit (or group of fields). The water losses from the source to the unit are not estimated. Nevertheless, these losses in surface water irrigation are significant.

Other studies have calculated the volumes used in surface water irrigation as the water actually taken from the reservoirs. For example, the White Paper on water in Spain (MIMAM, 2000) estimates average consumption of

Table 5.3. Comparison of irrigation using surface and groundwater in Andalucía

Indicator for irrigation	Origin of irrigation water			Ratio groundwater / surface water
	Groundwater	Surface water	Combined	
Irrigated surface (10^3 ha)[1]	210	600	810	0.35
Average volume applied in field (m^3/ha)[1]	4.000	5.000	4.700	0.8
Specific production (10^3 pta/ha)[1]	1.500	550	800	2.7
Total production (10^9 pta)	300	325	625	0.9
Water productivity (pta/m^3)[1]	360	110	160	3.3
Employment generated (UTA/100 ha)[1]	23	13	15	1.8
Total employment (10^3 UTA)*	50	75	125	0.67
Average consumption at origin (m^3/ha)	4.000[1]	7.400[3]	6.500[2]	0.54
Water productivity (pta/m^3)[3]	360	70	120	5.1
Employment generated (UTA/10^6 m^3)[3]	58	17	25	3.4

[1] Corominas (1999). Data relate to average volumes applied on the field.
[2] MIMAM (2000). Data relate to average volumes taken from reservoirs.
[3] Calculated using data from Corominas (1999) and MIMAM (2000), relating to average volumes taken from origin. (All data rounded up.)
* UTA stands for Working Units/Year (Unidades de Trabajo-Año) which is the work of one person working full-time for one year.

6700 m^3/ha/year and 6500 m^3/ha/year for the same geographic areas that are the subject of the Andalucía study, and without differentiating between surface and groundwater irrigation. Using these new figures and the volumes given for irrigation with groundwater in the Andalucía study, we can estimate more realistic average volumes used for irrigation with surface water of 7400 m^3/ha/year. Using these new more realistic data, the lower part of Table 5.3 shows that the productivity of groundwater irrigation is five times greater than that using surface water and generates more than three times the employment per m^3 used.

It can be argued that the greater socio-economic productivity of groundwater irrigation in Andalucía can be attributed to the excellent climatic conditions that occur in the coastal areas. While good climatic conditions may influence the results, the situation is similar in other regions of Spain. A recent doctoral thesis from the University of Zaragoza (Arrojo, 2001) has shown similar results to those obtained in Andalucía. It compared two irrigated areas in Aragón (northern Spain), one using surface water and other using groundwater, and both having comparable climatic and basic socio-economic and technical conditions.

Although we have not found other comparable studies in Europe, studies in India also suggest similar results. For example, Dains and Pawar (1987) estimate that groundwater constitutes 30% of all water used for irrigated agriculture in India but is responsible for 70–80% of all agricultural production.

When analysing the data included in this section, it is important to keep in mind the strong uncertainties that are attached to hydrological data. However, the results obtained are indicative of the greater productivity of irrigation using groundwater. This should not be attributed to any intrinsic groundwater quality, but to the greater control and supply guarantee that groundwater offers, as well as to the fact that farmers have an incentive to use groundwater more efficiently, because they bear the full costs of drilling, pumping and distribution (see Llamas *et al.*, 2001, chapter 7).

Hydrological benefits

Another potential benefit from groundwater development is an increase in net recharge in those aquifers that, under natural conditions, have a water level close to the land surface. A draw-down of the water table can result in a decrease in evapotranspiration, an increase in the recharge from precipitation that was rejected under natural conditions, and an increase in indirect recharge from surface water bodies. This process was already described by the American hydrogeologist Theiss in 1940, and was later expanded by Bredehoeft *et al.* (1982).

A clear example of this situation is the increase in available resources for consumption that followed intensive groundwater pumping in the Upper Guadiana Basin in central Spain. Cruces *et al.* (1998) have estimated that average renewable resources may have increased by one-third under disturbed conditions. Figure 5.1 illustrates some results from this work. Prior to the 1970s, groundwater pumping in the Guadiana Basin did not have significant impacts on the hydrologic cycle. Intensive pumping for irrigated agriculture started in the early 1970s and reached a peak in the late 1980s. As a result, wetlands that under semi-natural conditions had a total area of about 17 000 ha today only cover 7000 ha. In addition, some rivers and streams that were naturally fed by the aquifers have now become net losing rivers. Figure 5.2 illustrates the evolution of the water table in the main aquifer in the system between 1975 and 1999, a period during which groundwater development has been extremely intense. It can be seen that during the last 3 years (1996–1999), which were humid, the water table recovered almost 50% of the previous depletion. This figure shows the need to evaluate the influence of groundwater withdrawals and decrease in natural recharge (as a result of decreased precipitation) during dry spells in conjunction with each other.

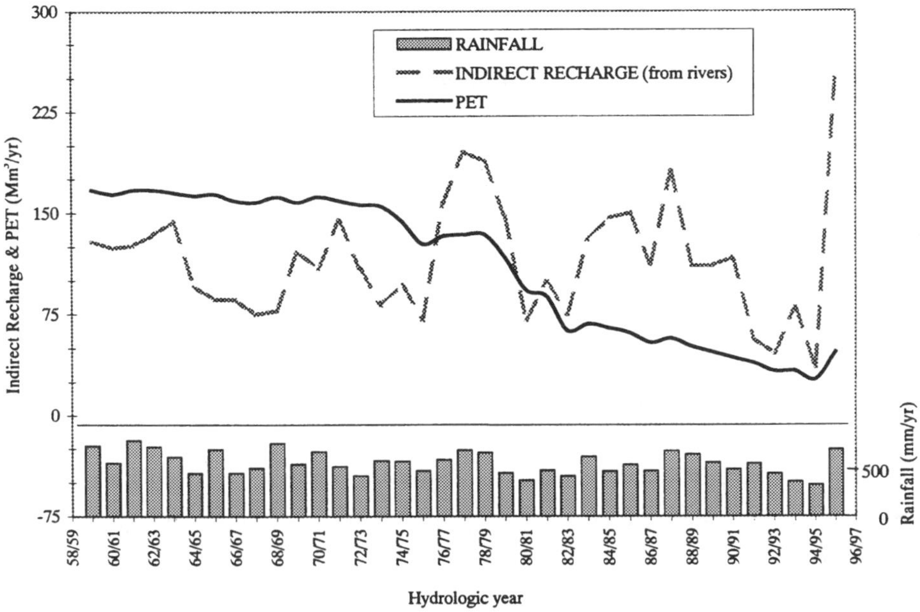

Figure 5.1. Indirect recharge from rivers and potential evapotranspiration from the water table and wetlands: evolution in the Upper Guadiana Basin aquifers. *Source*: Martínez Cortina (2001).

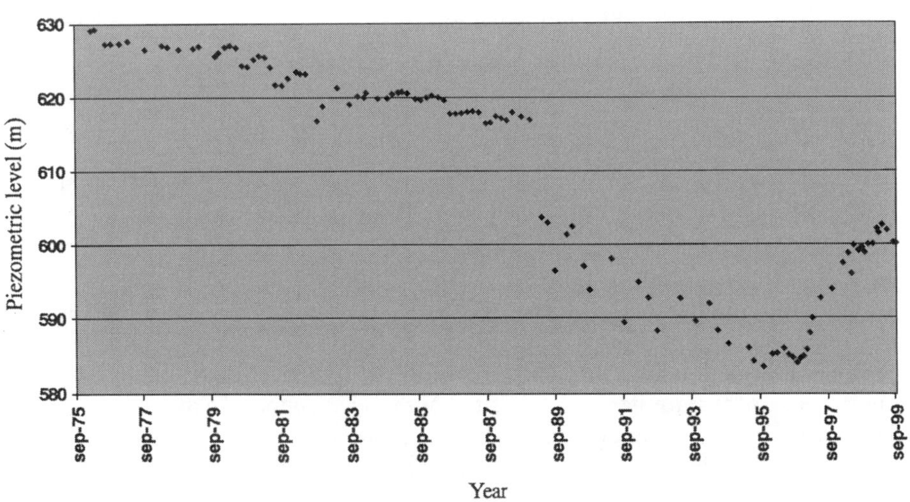

Figure 5.2. Piezometric level evolution in Manzanares (H.U. 04.04). *Source*: Martínez Cortina (2000).

The results of the draw-downs in the water table have been two-fold. On one hand, there has been a significant decrease in evapo-transpiration from wetlands and the water table, from about 175 Mm^3/year under semi-natural conditions to less than 50 Mm^3/year today. At the same time, there has been a significant increase in indirect recharge to the aquifers from rivers and other surface water bodies. This becomes particularly clear when looking at the last hydrologic year included in the model (1996/97), which was the first humid year after a long dry spell. While precipitation in this year was lower than precipitation in the 1960s (humid period in undisturbed conditions), recharge from surface water bodies was almost double due to the changes in the landscape. Consequently, more resources have become available for other uses. It is clearly important to keep in mind the associated negative impacts that the draw-down of the water table has had on dependant natural ecosystems.

5.3.2. *Groundwater level declines*

When a trend towards continuous significant decline in groundwater levels is observed, this is frequently considered to be an indicator of imbalance between abstraction and recharge. While this may be the case, the approach may be somewhat simplistic. When overall output from an aquifer is greater than input, the difference comes from groundwater storage and water levels fall. But a reversal in this trend may take some time to become apparent, since it will depend on aquifer characteristics such as size, transmissivity, degree of stratification and heterogeneity (Custodio, 1992; Adams and MacDonald, 1995). Management decisions based on this simplistic approach may sometimes be misguided. For instance, in large unconfined aquifers, the time necessary to reach a new state of equilibrium in water table levels can be decades or centuries when transmissivity is very low (Custodio, 1992, 1993; Bredehoeft *et al.*, 1982). On the other hand, in large confined aquifers, water level declines do not necessarily imply a significant decline in storage but rather a change in the elastic conditions of the system.

With respect to the climatic sequence, in arid and semi-arid countries, significant recharge can occur only every 5 to 10 years. Continuous decline in the water table during a dry climatic sequence, when recharge is low and abstraction high, may therefore not be representative of long-term trends. Declines in water levels should indicate the need for further analysis. Whether they indicate possible over-exploitation is something that needs to be studied on a case-by-case basis, always taking into consideration the hydrogeological characteristics and the size of the aquifer, as well as the climatic sequence. In any case, whether over-exploitation (in the sense of abstraction greater than recharge) is occurring or not, declines in the water table can result in a decrease in the production of wells as well as increases in pumping costs. This economic impact can be more or less significant depending on the value of the crops obtained. For instance, in some zones of Andalucía, the value of crops in greenhouses may reach 40 000–60 000 US$/ha/year. The water volume used is between 4000 and 6000 m^3/ha. The energy needed to pump 1 m^3 100 m higher is about 0.3 kwh. This means that the increase in the costs of pumping in the event of a drawdown of 100 m is almost irrelevant for the agribusiness. On the other hand, if the value of the crops is only about 1000 US$/ha/year, and the water needed is about 10 000 m^3/ha, obviously the increase in the cost of energy of a drawdown of 100 m can make that agriculture economically unfeasible.

5.3.3. *Degradation of groundwater quality*

Groundwater quality is perhaps the most significant challenge to the long-term sustainability of groundwater resources. Reclamation of contaminated aquifers can be a very costly and difficult task.

Most often, degradation of groundwater quality is not a result of excessive abstraction, but is related to other causes such as point or non-point source pollution from sources such as return flows from irrigation or leakage from septic tanks and landfills. However, groundwater abstraction can also cause changes in groundwater quality. Some indicators of the susceptibility of an aquifer to water quality degradation are the following:

- Proximity to saline water bodies: risk of saltwater intrusion which not only depends on the amount of abstraction relative to recharge, but also on the well field location and well design, and on the geometry and hydrogeological parameters of the aquifer (Custodio, 1992, 1993).
- Hydraulic connection to low-quality surface or groundwater bodies. Changes in the hydraulic gradient as a result of groundwater abstraction may result in the intrusion of poor-quality water into the aquifer from adjacent water bodies.

In these cases, the problem is often related to inadequate well field location and not necessarily to the total volumes abstracted (Custodio, 1992, 1993). Technical solutions to deal with problems of saline or lower quality water intrusion have been developed and applied successfully in some places (for example, California and Israel) (Custodio and Bruggeman, 1987).

5.3.4. *Susceptibility to subsidence and/or collapse of the land surface*

Aquifers formed in young sedimentary formations are prone to compaction as a result of water abstraction and the resulting decrease in intergranular pore pressure. This has been the case in the aquifers underlying Venice and Mexico City, for example. More dramatic collapses are a common occurrence in karstic landscapes, where oscillations in the water table as a result of groundwater abstractions can precipitate karstic collapses. In both cases, the amount of subsidence or the probability of collapse is related to the decrease in water pressure. This is a result not only of the amount of groundwater withdrawal, but also of well field location and design.

5.3.5. *Interference with surface water bodies and streams*

Decreases in the water table as a result of groundwater withdrawals can affect the hydrologic systems of connected wetlands and streams. Loss of base flow to streams, desiccation of wetlands, and transformation of stretches of rivers that were previously gaining into losing, may all be potentially undesirable results of groundwater abstraction and serve as indicators of possible over-exploitation. The Upper Guadiana catchment in Spain is a typical example of this type of situation (Cruces *et al.*, 1998).

5.3.6. *Ecological impacts on groundwater-dependent ecosystems*

The ecological impacts of draw-downs of the water table on surface water bodies and streams are increasingly constraining new groundwater developments. Drying up of wetlands, disappearance of riparian vegetation because of decreased soil humidity, or alteration of natural hydraulic river systems can all be used as indicators of over-exploitation. Reliable data on the ecological consequences of these changes is not always available, and the social perception of such impacts varies in response to the cultural and economic situation of each region.

The lack of adequate scientific data to evaluate the impacts of groundwater abstraction on the hydrological systems of surface water bodies makes the design of adequate restoration plans difficult. For instance, wetland restoration programmes often ignore the need to simulate the natural hydrological regime of the wetland. That is, not only restoring their form but also their hydrological function. Similar problems result from trying to restore minimum low flows to rivers and streams. Minimum stream flows are often determined as a percentage of average flows, without emulating natural seasonal and year-to-year fluctuations to which native organisms are adapted.

The social perceptions of the ecological impacts of groundwater abstraction may differ from region to region and result in very different management responses. A recent European Union-funded project looked at the effects of intensive groundwater pumping in three different areas in Greece, Great Britain and Spain (Acreman *et al.*, 2000). In the Pang River in Britain, pressure by conservation groups and neighbourhood associations with an interest in conserving the environmental and amenity values of the river, which had been affected by groundwater abstraction, primarily drove the decisions made by management. In the Upper Guadiana Basin, dramatic draw-downs in the water table (30–40 m) caused jointly by groundwater abstraction and drought (see Figure 5.2) resulted in intense conflict between nature conservation officials and environmental NGOs, irrigation farmers and water authority officials. This conflict has been ongoing for the past 20 years and has not yet been resolved.

Table 5.4. Institutional robustness indicators

Indicator	Measure
Regulatory framework	Strong Medium Weak
Monitoring and control networks	Dense Average Insufficient
Enforcement mechanisms	Strong Medium Weak
Implementation of grassroots organizations	High Medium Low

Source: Acreman *et al.* (2000).

Management attempts to mitigate the impact of water level drops on the area's wetlands have so far had mixed results. On the other hand, in the Messara Valley in Greece, the wetland degradation caused by drops in the water table has not generated any social conflict.

5.3.7. *Institutional robustness*

A final but crucial indicator of the susceptibility of an aquifer to over-exploitation is the ability of the institutional framework in place to respond to situations of stress. Possible indicators of this institutional robustness are suggested in Table 5.4. These indicators are subjective, given the difficulty of assigning precise numerical values to each one of them. However, they can serve to inform management decisions.

In cases where grassroots organizations are in place and adequately organized, it is possible to successfully implement a management plan to mitigate the undesired effects of excessive or poorly planned groundwater abstraction. In other cases, however, dealing with situations of stress may be more challenging. In Spain, for example, the 1985 Water Act has still not been successfully implemented in areas pertaining to groundwater management. Monitoring, control and enforcement mechanisms are insufficient and sometimes ineffective, so that, in situations of stress, they have been unable to respond adequately. The institutional framework is therefore not robust enough to organize a timely response, and the situation tends to worsen.

5.4. Groundwater use: renewable resources versus groundwater mining

It would seem necessary to make a distinction between groundwater over-exploitation and groundwater mining. The first concept is associated with the concept of renewable resources and has been the subject of this paper. Groundwater mining, on the other hand, occurs in areas where recharge is small or non-existent given low precipitation conditions. This is likely to be the case in some aquifers in the Middle and Near East and North Africa. A discussion about sustainable resource use in these regions must use other points of references than those used in this chapter. The discussion should primarily be an ethical one (Llamas, 1999).

In this sense, some authors consider that the traditional view that arid countries should develop in relation to their renewable water resources is erroneous (Lloyd, 1997). In their view, the ethics of the sustainability of non-renewable water resources should be considered in terms of continuous technological improvements. With adequate management, many arid countries could use their resources beyond the foreseeable future.

An important concept in this discussion is the idea of 'virtual water' suggested by Zehnder (1999). Zehnder argues that in arid countries, the difference between available water resources (both from renewable and non-renewable sources) and total resources needed is made up in the form of virtual water or imported food. In Israel, for instance, virtual water amounts to 60% of total water used. In Libya, the country using the largest proportion of fossil (non-renewable) water, available renewable reserves amount to only 110 m^3/inhabitant/year whereas those of fossil water are 770 m^3/inhabitant/year. The remaining 600 m^3/inhabitant/year that are needed to meet Libya's water demands are imported in the form of food.

In our view, groundwater mining could be a rational and ethical option if the following conditions are met:

- A correct hydrogeological evaluation has been done guaranteeing that at the projected rate of abstraction, the aquifer can supply water over a long term (for example, for 50–200 years, the projected life-span of many north African reservoirs).
- The ecological impacts and the economic viability of the project have been evaluated, including aspects such as the effects on other countries sharing the same resource.
- The project beneficiaries have been informed that the resource will eventually be depleted.
- Alternative water supply sources are planned.

5.5. Policy implications

5.5.1. *The concept of over-exploitation and the law*

The concept of over-exploitation understood exclusively in terms of abstraction rates superior to average long-term recharge rates has frequently been the basis for groundwater legislation and management decisions. In Spain, for instance, in the Regulation for the Public Water Domain that developed the 1985 Water Act, it is stated that "an aquifer is overexploited or in risk of being overexploited, when the sustainability of existing uses is in immediate threat as a consequence of abstraction being greater or very close to the mean annual volume of renewable resources, or when it may produce a serious water quality deterioration". Sixteen aquifers have been legally declared either provisionally or definitively overexploited. Strict legal regulatory measures have been designed and implemented to deal with these situations of stress. However, to a large extent these measures have not been successful and a situation of chaos still persists in many of these aquifers, as is implicitly recognized in the White Paper on water in Spain (MIMAM, 2000).

While the concept of over-exploitation may be useful, it is misguided to apply it strictly. We have seen how many other criteria, not merely hydrological ones, should serve to evaluate the level of stress in an aquifer. We have reviewed the issues of uncertainties, social preferences or institutional parameters that, in addition to purely hydrogeological information, influence the degree of fragility or susceptibility of an aquifer to stress. In this context, we suggest that a new, much broader approach be taken to deal with situations of scarcity and intense exploitation. A useful point of reference might be the Dutch model.

While for a long time a technocratic approach was used in the Netherlands to determine the amount of available groundwater resources (as a percentage of precipitation recharge), a much broader approach is currently used (Kors *et al.*, 1996, cited in GRAPES, 1998). Available resources are now estimated on the basis of a regional assessment of the adverse effects of conjunctive or independent use of surface and groundwater resources on the public water sector, public health, agriculture, nature reserves and so on. The concept of available resources is not strictly hydrological, but rather a dynamic concept that varies in response to new hydrological investigations, national and regional policy goals and changing social preferences. In this way, socio-economic, environmental and even institutional variables are taken into consideration and management responses can be designed accordingly. Maybe more importantly, the amount of renewable resources is a dynamic concept that can change, adapting to the changes brought about by human actions, management decisions and other variables.

5.5.2. *Policy recommendations for a renewed water resources policy*

A dynamic management model requires a more flexible approach to creating water resource policy. While regulation has been the primary tool used for resource allocation in southern European countries, it is no longer likely to serve current social demands on water resources, particularly in situations of scarcity and stress. Some limitations of a purely regulatory approach to water management are the following:

- A rigid licensing system where permits are granted for specific uses in specific

locations does not provide the necessary flexibility to adequately address situations of stress. A more flexible regulatory framework would allow for temporary or permanent transfers of water rights among users so that efficiency and equity criteria could be met. This greater flexibility is one of the goals of the Amendments to the 1985 Spanish Water Act, enacted in 1999.
- The effective implementation of regulatory tools requires the existence of adequate enforcement and control mechanisms to guarantee compliance with existing regulations. This is not always the case. In Spain, for instance, the existing legal framework gives river basin authorities extensive regulatory powers to deal with situations of aquifer over-exploitation. However, basin authorities do not have adequate enforcement and control capabilities so that non-compliance with regulatory restrictions is rampant.
- A water management system based primarily on regulatory means also requires the existence of reliable and generally accepted information on existing resources and existing uses. As we have previously seen, in many southern European countries, data are scarce and/or of poor quality. When adequate information is not available, more participatory management tools are required to guarantee social acceptance and effective implementation of management programmes (Hernández-Mora, 1998).

In view of these limitations, the new approach outlined above requires greater user participation in management decisions, which implies significant educational efforts, and the search for alternative mechanisms to deal with situations of conflict. A necessary starting point for any adaptive and participatory management strategy is, however, the need to invest significant resources in improving the quality of data.

Improving existing data

It would be necessary to do this on two fronts. Firstly, it is of key importance to make accurate inventories of existing licenses or rights of usage, as well as to continuously update such inventories. While this might seem self-evident, conflicts over existing rights too often hamper the effective implementation of management plans. In Spain, for instance, the process of making inventories of existing rights of usage in order to calculate total abstraction volumes has been ongoing since 1985 and is still uncompleted and a source of conflict.

The second and equally important need is to improve the quality and the quantity of hydrogeological data. It is to be hoped that the EU Water Framework Directive contributes to harmonize hydrological monitoring efforts in member countries.

Information and education programmes

Given the complexity of groundwater resources management issues, effective and informed public participation can only take place if there is a concerted effort to educate the public. It is necessary that the public be aware of the implications of their choices and it is the responsibility of public resource management agencies to provide comprehensive information and the skills to understand it.

With respect to information, the European Union's Directive 90/313 on facility of access to environmental data has established the legal basis for transparency and openness in natural resources management in member countries. However, this measure alone is not sufficient. Information needs to be taken to the public and presented in a form that is both easily understandable and easily accessible. Use of the Internet to disseminate information is one possible route. For example, the United States Geological Survey posts continuously updated information easily accessible to interested parties on stream flow, water quality, weather, and other hydrological data. It is, however, important to keep in mind that Internet access is still limited to a minority of the population in many EU countries. Resource managers therefore need to publicize management proposals, environmental data and other issues through more conventional means such as public information and discussion meetings in affected areas, consultation with non-governmental organizations, demonstration projects, interpretative centers, publication of brochures, etc.

Public education programmes are also important ways of disseminating information and informing public decisions. Educational activities should range from formal educational programmes in primary and secondary schools to continuing education of the public-at-large. They should also range from providing general information on hydrological issues, to specific information and education campaigns relating to specific projects, strategic planning programmes and so on. Some initiatives that could be taken are:

- preparation of posters, brochures, booklets and educational videos both on general water issues (hydrological cycles) and on specific projects or proposals;
- workshops for primary and secondary school teachers on groundwater resources education;
- workshops and short courses for leaders and members of different stakeholder groups;
- workshops and meetings of scientists and resource managers to exchange information on new technologies, innovative management approaches, etc.;
- round table discussions with the participation of members from different stakeholder groups to learn from and discuss each other's point of view.

Stakeholder participation in decision-making processes

Given the limitations of using a strictly regulatory approach to allocating existing resources between competing uses, increasing participation of current and potential stakeholders in all stages of design and implementation of management plans is essential. This may be particularly necessary in times of stress.

While public participation in water management is not a new concept, it is often limited to certain interest groups or to certain stages of the management process so that participation is not truly effective. In general, it is possible to distinguish between three phases of evolution in public participation programmes. Different countries are at different stages in this evolution.

At first, public participation is understood in a very limited sense, as a need to educate and inform the public of management decisions. This is not true participation, but rather a unilateral communication where the public is informed about decisions already taken and in the process of implementation. The public has therefore no opportunity to influence the decision making process. The information is at the disposal of the public, but no efforts are made to take the information to them or disseminate it, or to gather feedback from interested parties.

Second stage communication between management agencies and the public is more fluid and is two-directional. Public opinion can, to a certain extent, influence management decisions. While the process is more participatory, it is still the public sector that controls the decision making process. The Environmental Impact Statement laws that are in effect in European Union countries have brought most countries to this second stage.

However, true participation only occurs in a third stage, in which management agencies move from informing the public and receiving their opinions to actually deciding with the public (Delli Priscolli, 1998). The effort required from management agencies is significant, as are the possible resulting risks. In this sense, once all stakeholders have had a chance to express their views and interests, the challenge then becomes how to reconcile often contradictory positions while keeping everyone involved and satisfied. It is at this point that it becomes necessary to design conflict resolution mechanisms with the goal of reaching solutions that are acceptable to all. While the process becomes more consuming of effort and time, implementation of the mutually agreed plans will be significantly easier and has the greatest chance of success.

5.6. Final remarks

Complexity and variability characterize water management problems in general, and even more so in the case of groundwater. Uncertainty is an integral part of water management. This uncertainty is related to the scarcity of data, strong non-linearities in groundwater recharge values and changing social preferences. Honesty and prudence in recognizing current uncertainties is necessary. At the same time, there needs to be a concerted effort to obtain more and better hydrological data to inform management decisions.

Aquifer over-exploitation is a complex concept that needs to be understood in terms of a comparison of the social, economic, and environmental benefits and costs that derive from a certain level of water abstraction. It is difficult to define over-exploitation in purely hydrogeological terms given uncertainties in recharge and abstraction values and the fact that the amount of available resources in a catchment is variable and can be influenced by human actions and management decisions. The assumption that there is a trend towards long-term decline in groundwater levels (over 10–20 years, for example), suggesting real over-exploitation or overdraft, may be too simplistic and misleading.

Increasing emphasis on cost-effective and environmentally sensitive management practices places a new emphasis on broad public involvement in any water management decision-making process. But guaranteeing effective public participation in management processes requires informing and educating the public on increasingly complex scientific and technical issues. Effective information and education campaigns are therefore essential. The conflicts that are often a part of water management processes require the use of innovative conflict resolution mechanisms that will allow for the discovery of feasible solutions that are accepted by all and can be successfully implemented.

Note

1. The study's database and the maps and tables used are accessible on the Internet at http://www.cap.junta-andalucía.es, under the heading "Inventario de Regadíos".

References

Acreman, M., Almagro, J., Álvarez, J., Bouraoui, F., Bradford, R., Bromley, J., Croke, B., Crooks, S., Cruces, J., Dolz, J., Dunbar, M., Estrela, T., Fernández-Carrasco, P., Fornés, J., Gustard, A., Haverkamp, R., de la Hera, A., Hernández-Mora, N., Llamas, M.R., Martínez Cortina, L., Papamastorakis, J., Ragab, R., Sánchez, M., Vardavas, I., Webb, T., 2000, Technical Report to the European Union ENV4-CT95-0186 – Groundwater and River Resources Action Programme on a European Scale (GRAPES), Institute of Hydrology, Wallingford, Reino Unido, 248 pp.

Adams, B., MacDonald, A., 1995, Over-exploited aquifer – final report, British Geological Survey, Technical Report WC/95/3, 53 pp.

Arrojo, P., 2001, Valoración de las aguas subterráneas en el marco económica general de la gestión de aguas en España, in: Hernández-Mora, N., Llamas, M.R., eds., La economía de las aguas subterráneas y su gestión colectiva. MundiPrensa y Fundación Marcelino Botín, Madrid (in press).

Bredehoeft, J.D., Papadopulos, S.S., Cooper, H.H., 1982, The water-budget myth: scientific basis of water management, Studies in Geophysics, National Academy of Sciences, pp. 51–57.

Corominas, J., 1999, Papel de las aguas subterráneas en los regadíos, Jornadas sobre las Aguas Subterráneas en el Libro Blanco del Agua en España, Asociación Internacional de Hidrogeólogos-Grupo Español, pp. 65–79.

Cruces de Abia, J., Fornes, J., Casado, M., Hera A. de la, Llamas, M.R., Martínez, L., 1998, El marco natural, agua y ecología, in: Cruces *et al.*, eds., De la Noria a la Bomba. Conflictos sociales y ambientales en la cuenca alta del río Guadiana, Editorial Bakeaz, Bilbao, pp. 17–130.

Custodio, E., 1992, Hydrogeological and hydrochemical aspects of aquifer over-exploitation, in: Simmers *et al.*, eds,, Selected Papers in Hydrogeology, International Association of Hydrogeologists, Heise, Hannover, vol. 3, pp. 3–28.

Custodio, E., 1993, Aquifer intensive exploitation and over-exploitation with respect to sustainable development, Proceedings of the International Conference on Environmental Pollution, European Centre for Pollution Research, vol. 2, pp.509–516.

Custodio, E., Bruggeman, K.A., 1987, Groundwater problems in coastal areas, Studies and Reports in Hydrology, No. 45, UNESCO, Paris, 576 pp.

Dains, S.R., Pawar, J.R., 1987, Economic returns to irrigation in India, New Delhi, Report prepared by SDR Research Groups Inc. for the US Agency for International Development.

Delli Priscolli, J., 1998, Public involvement, conflict management and dispute resolution in water resources and environmental decision making, in: Public Involvement and Dispute Resolution, Institute for Water Resources. U.S. Army Corps of Engineers, Virginia. pp. 45–62.

European Environmental Agency, 1996, Groundwater monitoring in Europe, Topic Report 14/96. (http://themes.eea.eu.int/showpage.php/state/water).

Fetter, C.W., 1994, Applied Hydrogeology, MacMillan, New York.

GRAPES Project, 1998, Second Annual Report to the European Commission. A. Gustard, coord. Contract ENV4-CT 95-0186. UK, 200 pp.GRAPES Project, 2000, Technical Report. A. Gustard, coord. Contract ENV4-CT 95-0186. Preprint. 250 pp.

Hernández-Mora, N., 1998, Stakeholder participation in groundwater management in the Western Mancha aquifer: Opportunities in the face of conflict and scarcity, in: El Agua a Debate desde la Universidad: Hacia Una Nueva Cultura del Agua. Proceedings on CD-ROM, Congreso Ibérico Sobre Gestión y Planificación de Aguas. Zaragoza, Spain, 14–18 September, 1998.

Kors, A., Claessen, F., Vermulst, H., van Ek, R., Bos, H., Boven, E., de Lange, W., Arnold, G., 1996, Water System Reconnaissance 1996. Policy Analysis Water Systems Reconnaissance. Theme: Droughts and groundwater. RIZA Nota 97.041, Lelystad (in Dutch).

Llamas, M.R., 1998, Over-exploitation of groundwater (including fossil aquifers), Proceedings of the UNESCO Congress: Water into the 21st Century: A Looming Crisis? Paris, 2–5 June, 1998, Vol. 2.

Llamas, M.R., 1999, Considerations of ethical issues in relation to groundwater development and/or mining, UNESCO International Conference on Aquifer Systems. Trípoli, 20–24 November 1996, 20 pp. (in press).

Llamas, M.R., Fornés, J., Hernández-Mora, N., Martínez Cortina, L., 2001, Aguas subterráneas: retos y oportunidades, Mundi-Prensa y Fundación Marcelino Botín.

Llamas, M.R., Back, W., Margat, J., 1992, Groundwater use: equilibrium between social benefits and environmental costs, Applied Hydrogeology, 1, 3–14.

Lloyd, J.W., 1997, The future use of aquifers in water resources management in arid areas, The Arabian Journal for Science and Engineering, 22, 33–45.

Martínez Cortina, L., 2001, Estudio mediante modelos numéricos de la recarga en grandes cuencas sedimentarias. Aplicación a la cuenca alta del Guadiana, Doctoral dissertation in progress. Universidad de Cantabria.

MIMAM, 2000, Libro Blanco del Agua en España, Secretaría de Estado para Aguas y Costas, Dirección General de Obras Hidráulicas y Calidad de las Aguas, Ministerio de Medio Ambiente, Madrid, 640 pp. approx.

MOPTMA (Ministerio de Obras Públicas, Transporte y Medio Ambiente), 1995, Libro Blanco de las Aguas Subterráneas. Centro de Publicaciones, MOPTMA. 135 pp.

Zehnder, A., 1999, Water use and food production – an international collaboration?, EAWAG News, 46E, June 1999.

PART III

Groundwater Quality Management: Regulation, Economic Instruments and Voluntary Approaches

CHAPTER 6

Controlling Groundwater Pollution from Agricultural Non-point Sources: an Overview of Policy Instruments

Cesare Dosi and Naomi Zeitouni

6.1. Introduction

Agricultural incidental impacts upon groundwater quality can be either traced back to the use of potentially harmful inputs, such as fertilizers and plant-protection products, or to other farming practices (irrigation techniques and groundwater abstractions). Aquifer enrichment takes place through pollutants accumulating on farmland (e.g. nitrogen surpluses) or coming from outside the farm-gates (saltwater in coastal areas) (Giacomelli *et al.*, Chapter 3, this volume).

Although the agricultural activities and the natural processes through which pollutants are generated and/or intruded into aquifers may be quite varied, they often share a number of general features. For instance, most of the agriculture-related groundwater pollution problems can be described as being non-point source (NPS) problems, in that they typically involve many geographically dispersed agents which cause intermittent low-pollution discharges which, in general, cannot be easily intercepted and neutralized through end-of-pipe structural devices. These features tend to make it difficult, or even impossible, to apply the battery of environmental policy measures traditionally employed to manage pollutant discharges from large and readily identifiable industrial and municipal point sources.

According to an EC Commission communication on the state of Europe's environment, whilst there have been substantial improvements in surface water quality due to reductions in point source discharges such as emissions of phosphorous (by 30/60% since the mid-1980s) and organic matter discharges (reduced by 50/80%), pollutant emissions from agricultural diffuse sources have shown little change since 1980, and EU maximum groundwater concentrations of nitrate and certain pesticides are frequently exceeded (European Commission, 1999b). In 1996, the Commission released the proposal entitled *An Action Programme for Integrated Groundwater Protection and Management*, where it is stressed that proper management of groundwater should be a key component of Member States' environmental policies, and, within the overall

C. Dosi (ed.), Agricultural Use of Groundwater, 129–155.

objective of groundwater conservation, "relieving the pressure from diffuse sources should have the highest priority" (European Commission, 1996).

There are various explanations for the still modest control of NPS pollution in general, and in particular, actions to regulate groundwater pollution from agricultural sources. One explanation surely lies in the difficulties still faced by policy-makers in updating traditional pollution control strategies and regulatory approaches in order to address NPS problems. For instance, water pollution control has mostly relied upon ex post structural correctives (privately or collectively managed water treatment facilities), or ex ante regulatory measures which take observable individual emissions (mandatory effluent standards or, less frequently, environmental charges) as a reference point. In the case of agricultural NPS pollution, due to the high cost of monitoring individual pollutant discharges, transaction costs[1] associated with regulatory policies are particularly high. "These higher costs may be one of the reason why point sources ... have been emphasised in water quality legislation" (McCann and Easter, 1999, p. 402).

Secondly, economic activities that are responsible for NPS pollution problems, agriculture in particular, have substantially been, and are still, although to a lesser extent, exonerated from mandatory regulation, or have not been confronted by effective economic incentives aimed at internalizing the social costs (benefits) of pollution (abatement). On the contrary, rather than addressing market failures and promoting a more sustainable use of natural resources, agricultural policies have often added further distortions, and by so doing, have often worsened the misuse of resources.

The purpose of this chapter is to provide a taxonomy and literature review of policy instruments for controlling pollution from agricultural diffuse sources, and compare their pros and cons.[2] The review also includes a description of instruments introduced though recent reforms of the European Common Agricultural Policy (CAP), reforms aimed, inter alia, at integrating environmental protection into policies traditionally designed to achieve other objectives.

6.2. Groundwater pollution from agricultural non-point sources: key features and implications for policy design

6.2.1. *NPS pollution: key features*

The underlying characteristics of NPS pollution have been documented in studies dating from the late 1970s. These features can be summarized as follows.

First, it is difficult to rely on structural devices for intercepting and neutralizing polluting substances. For instance, while discharges of waste water from industrial plants or municipal point sources are generally easy to treat, for example, by installing filters in the pipes through which effluents are released in the environment, NPS effluents are difficult to intercept and neutralize because of the geographical dispersion of sources, and because pollutants may follow tortuous paths before reaching water bodies.

The second important feature of NPS pollution is the part played by the physical characteristics of the site where farmers operate, as well as the area through which pollutants move, in determining both the generation of potential pollutants ('on-site emissions') and their ultimate environmental effects ('off-site discharges'). For instance, the same farming practice may have different impacts upon water quality depending on the characteristics of the farmland, climatic conditions, and the location of the farm in relation to potentially affected water bodies. This aspect is not exclusive to NPS pollution. Even for point sources, the firm's type (e.g. the relative efficiency of machinery and equipment) and location may affect on-site emission rates and their ultimate environmental impacts. However, there are two features which tend to characterize NPS problems (Dosi and Tomasi, 1994). The first is the sheer number and variety of sources (heterogeneity of farmland characteristics, hydrological and climatic conditions). The second is the role played by exogenous and partly unforeseeable events (such as weather conditions) towards the generation of potential pollutants and their delivery ratio (i.e. the ratio between off-site discharges and on-site emissions).

The third and probably most definitive feature of NPS pollution is the difficulty of monitoring individual pollutant discharges. While pollutants from point sources enter the environment at a specific, single location (such as a single pipe), NPS effluents (which often have a fairly low density per unit area) do not enter water bodies at a defined point, and are usually dispersed by natural processes. Inferring individual responsibilities from observable ambient pollutant concentrations is also difficult. While pollutants from point sources are usually delivered to water bodies more or less proportionally to on-site emissions, NPS pollutants may travel long distances and undergo a qualitative change before delivery.

The underlying features of NPS pollution that have been described have two main implications for policy design. First, the difficulty of relying upon end-of-pipe structural correctives makes a preventive approach (abatement of on-site emissions) preferable; it is sometimes the only viable option for controlling water pollution from diffuse agricultural sources. Second, because of the difficulty/impossibility of monitoring individual discharges, the effectiveness of regulatory measures aimed at preventing the generation of pollutant loads depends essentially on policy-makers' (willingness and political) ability to enforce alternative ways of establishing the causal link between farmers' activities and observable groundwater quality problems.

6.2.2. *A taxonomy of NPS pollution control policy instruments*

General classifications of environmental policies
Before focusing on the classification of specific measures proposed to address NPS pollution problems, let us first briefly look at the more general taxonomies of environmental policy instruments.

A common classification is one that highlights the difference between economic instruments, voluntary approaches, and mandatory regulations. A standard definition of economic instruments can be found in OECD (1991, 1997), where they are described as "instruments that affect costs and benefits of alternative actions open to economic agents, with effect of influencing behaviour in a way that is favourable to the environment" (OECD, 1991, p. 10). These instruments typically involve either a financial transfer between polluters and the community (e.g. charges/taxes or subsidies) or the creation of new markets (e.g. tradable emission/pollution permits). So-called voluntary approaches (VAs) are somewhat more elusive and difficult to define. A quite general and comprehensive definition of VAs is provided by Lévêque (1997), who describes them as commitments of polluting firms or sectors to improve their environmental performance. According to Brau and Carraro (1999), these commitments can be placed into three categories: (i) unilateral commitments, which consist of environmentally friendly adjustments established by firms themselves (e.g. a spontaneous switch to organic farming, either for ideological reasons, or taking advantage of consumers' willingness to pay for green products); (ii) public voluntary schemes, in which participating firms agree to standards developed by public bodies (e.g. farmers' adhesion to the agri-environment schemes, introduced through Regulation 2078/92, described in section 3); (iii) negotiated agreements, i.e. specific contracts between public authorities, or other intermediate subjects, and polluting firms, e.g. agreements between water authorities, or water supply companies, and farmers operating within or near drinking water catchment areas (see Chapters 7, 8 and 9, this volume). Finally, policy provisions which do not make appeal to economic rationality or social responsibility, but involve a compulsory restriction of the polluters' choice domain, can be labelled command and control policies.

Besides the distinction between policies aimed at promoting self-regulation and mandatory regulations, a complementary criterion for classifying environmental policies is whether or not the polluter pays principle (PPP) applies. Broadly speaking, policy instruments are generally believed to be consistent with PPP if agents who use the environment either deliberately or incidentally as a sink for pollutants face a cost for the damage imposed on the rest of society. However, labelling policy provisions according to their consistency with PPP becomes more difficult when compensation is foreseen for economic agents who voluntarily commit themselves to go beyond (overcomply) the minimum environmental standards set up by mandatory regulations. Whilst these payments may not, at first glance, appear to be consistent with the PPP ethics, they are claimed to be so in various official documents in which over-compliance is ex lege assimilated to the provision of environmental services.

In this respect, it is worth noting that the European Commission has quite explicitly identified the legal borderline between negative environmental externalities, whose internalization does not make farmers eligible for compensation, and the provision of environmental services which, on the contrary, should be remunerated by society. The underlying rationale of the Commission's proposals for integrating environmental concerns into agriculture rests on two principles:

- first, farming, as any economic sector, should attain a basic standard of environmental care without specific payment. This should be contained within the scope of good farming practice (which includes many matters other than environment) and comprises observance of regulatory standards and an exercise of care which a reasonable farmer would employ. This basic standard is also referred to as the reference level;
- second, wherever society asks farmers to provide an environmental service beyond the reference level, and the farmer incurs cost or income loss, society must expect to pay for the service. This standard is also known as the target level (European Commission, 1998, p. 115).

A taxonomy of NPS pollution regulatory strategies: indirect and direct policy approaches

There is a large body of literature dealing with policy instruments aimed at controlling water pollution from agricultural diffuse sources through preventive measures. The proposed instruments can be classified according to the criteria given above. One classification is the way in which pollution control operates, i.e. through introducing compulsion to the farmers' choice domain or through affecting the pros and cons of alternative courses of action legally open to farmers. Another classification is the social distribution of the costs of pollution abatement (i.e. whether farmers are compensated for environmentally friendly adjustments).

However, when considering policies specifically addressing NPS pollution, it is useful to adopt an additional classification criterion based upon the way in which the monitoring problems that arise from the characteristics of NPS pollution are addressed. For instance, whilst most of the authors base their recommendations on the difficulty of monitoring individual impacts upon water quality, the proposed regulatory approaches vary across the economic literature.

Following the taxonomy proposed by Dosi and Moretto (1993), regulatory approaches can be classified according to the reference basis adopted for setting policy measures, namely *estimated* individual pollutant discharges (indirect regulatory approach) or *observable* total discharges (direct regulatory approach).

As far as the indirect approach is concerned, from the seminal papers of Griffin and Bromley (1982) and Shortle and Dunn (1986) onwards, many authors have considered that estimated rather than observable individual impacts upon water quality should provide the point of reference for designing regulatory tools (economic instruments or mandatory regulations). Such estimates could be obtained by means of available, albeit imperfect, models of pollutant generation (and transport) which provide predictions of on-site emissions (off-site discharges) attributable to a single farm or to a specific set of farming practices (Giacomelli *et al.*, Chapter 3, this volume). For example, methods for calculating nitrogen surpluses (known as the farm-gate balances) have been developed in some Member States in order to highlight areas at risk of nitrogen pollution (European Commission, 1999a).

In contrast, there are authors who have recommended observed concentrations of pollutants at particular water bodies (e.g. nitrate concentration in a confined aquifer) as an alternative to estimated individual impacts upon water quality, The rationale behind the direct regulatory approach is that by setting an incentive mechanism based on an observable variable (total off-site discharges) the regulator would induce certain unobservable actions (abatement of individual on-site emissions). Policy instruments consistent with such a regulatory approach typically take on the form of tax/subsidy schemes that, broadly speaking, depend on deviations between measured and desired ambient pollutant concentrations.

6.3. NPS pollution control: the indirect regulatory approach

6.3.1. *Estimated emission charges and standards*

Emission charges

Environmental charges may be considered as being a way of putting prices on the use of the assimilative capacity of the environment. In practice, they work either as emission charges or as product charges. The former are charges on effluents, and the tax burden is calculated according to the quantity or quality of pollutant emissions; the latter are levied on products (raw materials, intermediate or final products) whose quantity consumed or produced is taken as a proxy of the ultimate environmental impacts of a specific economic activity. In general, emission charges are considered to be more efficient than product charges because they leave target agents the freedom to select more cost-effective strategies to reduce effluents.

Obviously, the environmental effectiveness of an effluent-charge system crucially depends on the regulator's technical and administrative ability to monitor target agents' emissions and to evaluate their ultimate environmental impacts (pollutant concentrations at the receptor water body). Because of the number and geographical dispersion of sources, the often intermittent nature of pollutant emissions, and the spatial variability of transfer coefficients between on-site emissions and pollutant concentrations at the receptor, it follows that a charge system based upon observable individual discharges is not, in general, a viable policy option for dealing with NPS pollution problems. Similar considerations apply to reward systems foreseeing polluters' subsidization according to observable reductions of effluent-discharges.

However, as advocates of the indirect regulatory approach emphasize, the difficulty of monitoring individual emissions could be partly overcame if estimated, rather than observable individual pollutant discharges, form the basis of the system.

The decision to adopt individual estimated discharges as a reference basis for implementing economic instruments aimed at affecting polluters' behaviours has two main policy implications.

First, the model used to estimate individual environmental impacts has to be granted regulatory legitimacy, i.e. it has to be defined as the legal basis for computing the tax burden (Dosi and Moretto, 1993). Second, the adoption of an estimated emission-charge system requires that the regulator provide target agents with two types of schedules, one that relates the tax burden to the estimated effluents and one that relates estimated emissions to specific farming practices. As Shortle and Dunn (1986) stress, since the two schedules link the farmer's choice of farming practice or practices (e.g. application rates of nitrogen fertilizers) to the tax, a charge system based on estimated emissions is, in essence, equivalent to a product charge system (e.g. nitrogen fertilizer levies).

Emission standards

Emission standards do not represent a viable policy option for dealing with NPS pollution problems for the same reasons which make a charge system based upon observable individual discharges an impracticable regulatory strategy. However, the difficulty/impossibility of implementing an effluent-based regulation could be overcome if estimated, rather than observable emissions, form the basis of the regulatory scheme. Again, this requires granting legal legitimacy to a predictive NPS pollution model, and implies that the evaluation of compliance (or illegal behaviour) with the standard will be based upon monitoring farming practices which may or may not involve discharges exceeding the legally imposed emission threshold.

As a charge system taking estimated emissions as a reference basis is equivalent of a product charge system, a regulatory scheme based upon emission standards is, in essence, equivalent to regulatory schemes based upon 'technological standards' (e.g. restrictions on input use or mandatory codes of good agronomic practices).

6.3.2. *Input- and output-oriented policy measures*

In light of the above arguments, when individual pollutant discharges are not technically monitorable at a reasonable cost, input/output oriented policy measures (product charges, subsidization of environmentally friendly production methods, or technological standards), i.e. policies aimed at discouraging, promoting or mandating specific farming practice or practices, can be considered a reasonable second-best regulatory approach.

Input/output oriented policy measures aimed at addressing pollution from agricultural sources exhibit a great deal of variety. They include input and output levies, mandatory restrictions on input use, codes of good agricultural practice, reforms of agricultural policies, contingent subsidies (cross-compliance measures), and compensation for abandonment of potentially polluting activities (set aside).

Product charges, mandatory restrictions on input use and application zones
In principle, product charges such as levies on specific potentially polluting inputs should induce farmers to adopt precision technologies: i.e. in order to lower the tax burden, reducing input by using appropriate application methods to increase efficiency and plant uptake.[3] The difficulty of collecting data on input use for individual farmers may lead to a charge system based on observed choice of technology (or choice of crop). However, to be environmentally effective and economically efficient, such a system requires fixed proportion production and pollution technologies (Shah *et al.*, 1993).

When output (y) and pollutant discharges (z) are produced though variable input (a) using one or several distinct technologies (i): i.e. $y = f(a, i)$ and $z = g(a, i)$, an indirect optimal control of pollution can be established by output taxes or through several rates of input taxes that vary according to technology i. If a farmer adopts a technology which has a higher input-use efficiency (e.g. lower nitrogen surpluses), s/he will be charged through a lower tax rate (on output and/or on inputs).

However, establishing such tax rates is difficult because of non-linearity and the need to collect data on output and input use according to farmers' technology. It is obviously simpler to design uniform output or input levies in the form of sales taxes, but such uniform tax rates are sub-optimal. When there is a significant technological heterogeneity, taxing output may be especially counterproductive for those cases where farmers have adopted precision technologies: they will have higher levels of outputs and will have to pay higher taxes.

In addition to taking technological heterogeneity into account, when environmental variability is of greater significance, product charges should also be spatially differentiated. In other words, as with technological heterogeneity, heterogeneity in environmental conditions makes such economic incentives lose much of their theoretical appeal.

Since product charges such as levies on fertilizers and pesticides can hardly be distinguished within the same market, mandatory regulation, which may be spatially differentiated, may be more appealing (Zeitouni, 1991; Goetz and Zilberman, 1995). Optimal spatially differentiated mandatory regulations may be deduced by taking into consideration the hydrological properties of groundwater resources, their directions and speed of flow, and aquifer accessibility. Although these considerations require profound knowledge of the local conditions, the data needed to feed simple models may be available, and they can provide guidance for identifying areas locally sensitive to pollution, and the relevant application zones (Goralic *et al*., 1979; Millon, 1987; Zeitouni, 1991).

When information about aquifer properties is not available to the regulator, the optimal level of applied polluting inputs (such as fertilizers or pesticides) could be deduced by applying the safety first approach to risk management (Roy, 1952). In relation to this, Lichtenberg and Zilberman (1987) have suggested that the establishment of water pollution control policies should minimize the cost of attaining those environmental quality objectives which have a certain degree of statistical reliability. This requires that the probability of exceeding the quality target does not exceed a pre-specified level. From this optimization a shadow price for the risk can be calculated. This shadow price can be interpreted as the marginal cost for increased safety.

Braden *et al*. (1989) expanded this approach further in developing a regional land management and input choice model to reduce the cost of reaching a water quality target with a certain degree of reliability. Their analysis emphasized the importance of modifying farming practices in environmentally sensitive areas, either by direct control or by appropriate economic incentives. Although it is somewhat different, specifying quality standards in terms of risk instead of in concentrations enables differentiability in standards according to the sensitivity of the area to which they are to be applied. The reason is that a certain pre-specified risk over a sensitive location may entail more restrictive regulation, while the same risk applied to less sensitive locations may indicate a less restrictive concentration of pollutants.

It should be noted however, that setting a uniform level of risk for all places may not minimize costs. Lichtenberg *et al*. (1989) compared water quality standards for reducing the risk of DBCP in groundwater in California. One of the things they found was that regional risk targets could be met with reduced costs by setting lower standards for rural wells than for urban wells, since rural wells serve a smaller population.

Codes of good agricultural practice and vulnerable zones: the EC Nitrate Directive

Broadly speaking, the term 'good agricultural practice' refers to farmland management and production methods able to prevent or to reduce environmental damage. In EC legislation, the term is more commonly applied to the regulation of nitrate pollution from diffuse sources, and in this context, it can be seen as being an application to agriculture of the concept of best environmental practice that is applied in industry.

In 1991, the EC Council adopted the Nitrate Directive (91/676/EEC). Its aim was to reduce water pollution by nitrates from agricultural sources and to prevent further pollution. According to the Directive, Member States (a) must establish codes of good agronomic practice to be implemented by farmers on a voluntary basis, and (b) must identify vulnerable zones within their territory and implement action programmes which should contain mandatory measures for agricultural practices. The Directive defines a vulnerable zone as an area where nitrate concentrations exceed, or are likely to exceed in the future, the maximum admissible concentration of 50 mg/l.

Member States should have: (a) implemented the Directive in their national legislations, established a code or codes of good agricultural practice, and designated vulnerable zones by December 1993, (b) introduced action programmes imposing compulsory restrictions of farming activity by December 1995.

In 1997, the first planned Commission report on the implementation of the Nitrate Directive was produced. The Commission noted that 6 years after its adoption, "the status of implementation in most Member States is unsatisfactory [and] the failure to implement the Directive fully, in addition to its legal aspects, constitutes a failure to deal with serious environmental and human health problems" (European Commission, 1997). For instance, only four Member States met their implementation obligations by the set deadline (Denmark, Spain, France and Luxembourg). At the time the report was prepared, most Member States had yet to designate vulnerable zones (Belgium, Greece, Spain, Portugal and the UK). Action programmes, which should have started on December 1995, were notified to the Commission only by Germany, Luxembourg, Austria and Sweden, on June 1997.

Reforming agricultural policies: the EC 1992 agri-environment programme

Although pollution abatement generally requires the implementation of ad hoc environmental policies, in many instances polluters' behaviours could be positively affected through reforming existing sector policies to remove distorting incentives, or to integrate environmental protection into policies traditionally designed to achieve other public objectives. This is especially true for the European agricultural sector and for the Common Agricultural Policy (CAP).

As noticed by Brouwer (Chapter 13, this volume), it is difficult to assess the extent to which CAP has affected the course of agricultural development and, in particular, structural changes such as intensification, specialization, and concentration which are commonly believed to be responsible for observed negative water quality trends. For instance, even in a complete laissez faire scenario, European farmers would have overused the environment, because of the basic failure of market mechanisms to drive a socially efficient use of natural resources. However, one can legitimately claim that farming support policies, and in particular support though subsidization of commodity prices, rather than promoting a more efficient use of the environment, have often added further distortions (Dosi and Ferro, 1990).

For instance, at the time when CAP's objectives were drawn up, agricultural expansion (and expansion of production in general) was automatically accepted as being a desirable social goal, while environmental issues were considered to be extremely marginal. The heart of CAP was the system of guaranteed high prices for unlimited production which, by distorting output–input price relationships, has encouraged the intensification of agricultural activities and surpluses of farm products. Quotas on some products were introduced during the 1980s, but the purpose was to maintain guaranteed high prices, not to deliver, even indirectly, environmental benefits (European Environment Agency, 1995).

The legal requirement to integrate environmental protection into other EC policies was established in 1987 by the Single European Act (SEA) and was given a more comprehensive legal basis in the Maastricht Treaty. However, even before the SEA, the Commission acknowledged in various policy documents the need to update CAP in order to include environmental considerations. In particular, the 1985 *Green Paper Perspectives for the CAP*, stated explicitly that agriculture should be seen as being an economic sector which, like other sectors that are potentially damaging to the environment, should be subjected to restraints and controls in order to avoid environmental degradation, and that in general, the polluter pays principle should be applied (European Commission, 1985).

The need to inject substance into the general commitments made in the Green Paper, and other policy papers,[4] partly influenced the 1992 McSharry reform package, the first comprehensive and substantial update of CAP since the Treaty of Rome. This package included three measures to accompany the principal CAP reform measures, namely: (a) the agri-environment programme (Regulation 2078/92); (b) the early retirement scheme (Regulation 2079/92), and (c) the forestry aid scheme (Regulation 2080/92).

To properly interpret the environmental provisions included in the McSharry reform, it is worth recalling the surrounding political, budgetary and economic context. As Baldock and Lowe (1996) emphasize, "it would be wrong ... to see in this and subsequent policy initiatives the triumph of environmental interests .... Agricultural policy makers have responded to environmental concerns, not necessarily through any deep convictions, but because of the perceived coincidence between the aims of environmental improvement and the need to reduce agricultural output, thereby contributing to the alleviation of surplus and budgetary problems. [Moreover, especially in northern Europe] farming leaders, in a context of chronic oversupply of staple products and falling farm incomes, have begun to look to the provision by farmers of environmental 'products', in order to underpin or renew their claims for public support" (Baldock and Lowe, 1996, pp. 12–13).

As far as the potential contribution to water pollution abatement is concerned, the most important 1992 CAP reform accompanying measure is the agri-enviroment programme established through Regulation 2078/92,[5] which foresees compensations for farmers who undertake to reduce ('substantially') input use (namely fertilizers or plant protection products), to change to other more extensive crop patterns and more environmentally friendly production methods, or set aside farmland for at least 20 years with a view to protecting hydrological systems. While most of the Community farming support policies are not subject to additional funding by Member States, the agri-environment schemes are only partly financed though the European Agricultural Guidance and Guarantee Fund (EAGGF).

As recently reasserted in a Commission's report on the state of application of Regulation 2078, "[the agri-environment programme] is not a regulatory one and only intervenes in the range of activities over which a farmer has discretion to act. Thus action to prevent illegal pollution or to ensure that farmers observe minimum environmental standards in applying pesticides, should be the subject of regulation and codes of good agronomic practice. But not the aim of agri-environment measures." (European Commission, 1998, p. 18). Under Regulation 2078/92, the total expenditure by Member States for 1998 is estimated at ECU 1.73 billion, which represents about 4% of EAGGF which, in turn, accounts for about 50% of the entire Community budget.[6]

About 20% of the total European Union's farmland (EU15) has been affected by Regulation 2078, with significant differences, however, within and between Member States. For instance, in southern Europe (Greece, Spain, Italy, and Portugal), the percentage of hectares covered is below, sometimes well below, the average (0.6%, 2.9%, 13.6% and 16.8%, respectively). In France the percentage is slightly above the average (22.9%) (European Commission, 1998).

As far as the effectiveness of Regulation 2078 is concerned, according to the previously mentioned Commission's report, there is evidence of:

- "highly positive results ... for reduced input measures, especially organic farming, nature protection measures and maintenance of landscapes; some difficulties arose with extensification, set-aside for 20 years ... resulting in low take up".
- "arable conversion to extensive grass shows improvement in landscape quality in one region, while not enough data exists on reduction of N-leaching".
- "positive results from erosion prevention measures ... and N-leaching reduction measures, such as green-cover crops".
- "extensification of livestock measure has not been successful in several regions, one reason may be that the measures are not paid sufficiently".
- "application on highly profitable land is not satisfactory in the absence of sufficiently high premia. Greater use of targeting is generally suggested to ensure appropriateness of payments". (European Commission, 1998, pp. 7–8).

In general terms, according to the Commission, the results of the first agri-environment programme have been quite positive, in that "at 4% of CAP Guarantee spending, [the substantial environmental benefits] represent good value for money" (European Commission, 1998, p. 8).

However, besides the programme's internal rate of return, the key issue is whether or not only reliance upon farmers' voluntary undertaking of subsidized environmental friendly adjustments can be considered as being substantial progress toward integrating environmental objectives into CAP. In this respect, it is legitimate to state that Regulation 2078 has not been a very effective engine for driving widespread and substantial groundwater quality improvements.

As forecast by some commentators immediately after the approval of the McSharry package, the agri-environment schemes have proved to be not attractive, and, consequently, they have not significantly affected the behaviours of those farmers for whom the cost of abandoning environmental unfriendly farming practices is relatively high (following the terminology employed in the Commission's report, farmers operating on highly profitable land). For instance, operating on highly profitable land does not necessarily mean that there is a higher pressure upon groundwater quality. However, when there is an overlap between farmland productivity and environmental sensitivity, reliance upon voluntarism and untargeted subsidization is unlikely to be an environmentally effective (and efficient) policy provision. The shortcomings of a not properly targeted subsidization of environmentally friendly farming adjustment are testified to by the very modest impacts of Regulation 2078 upon intensive agricultural systems: in a large number of European regions, there have been little changes in groundwater pollutant concentrations.

Reforming agricultural policies: cross-compliance measures

Polluters can be induced to abandon certain practices or adopt certain conservation measures if this is set as a condition for eligibility for other public programmes that they find attractive. When these programmes foresee subsidization of output prices, the traditional and still prevailing CAP support scheme, cross-compliance measures can be interpreted as an implicit form of output charges, in that failure to acquire eligibility implies a reduction of (a levy on) guaranteed prices.

To our knowledge, cross-compliance measures tied to environmental objectives were first introduced in the USA, as part of the Conservation Title of the 1985 reauthorization of the Food Security Act. In the EC, they were only later formally considered as a policy option, and they were introduced though the recently agreed Agenda 2000, as a Member States' policy option.[7,8] The European Commission's position on the proper use of cross-compliance measures is quite clearly stated in various working documents: "cross-compliance is most appropriate in ensuring adherence to the *reference level*" (European Commission, 1998, p. 115),[9] i.e. attaining a basic standard of environmental care, and not the provision of additional environmental services involving costs or income losses that should be paid by society.

Generally speaking, the link between farming support (either in the form of price support or direct income support) and farmers' environmental performance can be implemented in different ways, ways that tend to exhibit a different degree of environmental effectiveness. Following the taxonomy proposed by Batie and Sappington (1986), two general approaches can be identified: (a) the red ticket approach, where eligibility for certain benefits (e.g. guaranteed prices) is made contingent upon the farmer attaining a given environmental standard or set of standards; (b) the green ticket approach, where farmers become eligible for higher levels of support if they comply with or exceed a given environmental standard.

It follows that the basic difference between the red and the green ticket approach is whether or not the benefits from existing farming support policy schemes are made contingent upon reduction of environmental damages, or whether pollution abatement per se entitles farmers to get additional benefits with respect to the farming support baseline. For instance, with a green ticket policy "a basic direct support is paid regardless of compliance with environmental standards and the additional support for complying or exceeding a given set of standards can be seen as a voluntary environmental scheme" (Christensen and Rygnestad, 1999, p. 5).

Christensen and Rygnestad (1999) provide examples of red and green schemes with reference to Danish legislation, which has implemented the Agenda 2000 reform. Reductions in hectare payments and headage payments are foreseen for farmers who do not complete field plans or fertilizer plans, and for farmers who do not complete fertilizer accounts and over-fertilize, respectively (examples of red ticket schemes). Farmers operating in designated areas must comply with certain farming practices, including a reduction of fertilizer use, in order to receive subsidies only provided for environmentally friendly agricultural practices undertaken in environmentally sensitive areas.

In between the red and the green ticket schemes, is what is described by Baldock (1993) as the orange ticket approach, where eligibility for support payments is dependent upon farmer's willingness to enrol in an otherwise voluntary scheme which attracts ad hoc payments. An example of an orange ticket policy is the US Conservation Reserve Program. The programme, introduced as part of the Conservation Title in the 1985 Food Security Act, was designed to achieve multiple objectives, namely conservation of soil resources, reduction of surplus stocks of agricultural products, enhancement of wildlife habitat, and maintenance of farm income (Dosi, 1994). Farmers were allowed to be included in the programme if at least one-third of cultivated fields were classified as highly erodible land, and strong penalties were established for violation of CRP contracts. These include loss of access to price support programmes, government crop insurance, loans, and, obviously, CRP payments.

Regardless of the 'colour' of the cross-compliance provisions, their environmental effectiveness (in terms of pollution abatement) obviously depends, first of all, on the attractiveness of the host programme (i.e. the programme providing the benefits which would be lost if a farmer fails to acquire eligibility) as well as on the cost of acquiring eligibility requirements. Cross-compliance measures are obviously pointless if farmers perceive that the cost of complying with pollution abatement requirements is higher than the foreseen reduction of benefits stemming from the host programme.

Moreover, the effectiveness of cross-compliance measures in terms of pollution abatement depends on the correlation between the economic characteristics of those farms which enrol in order to acquire eligibility, and the intrinsic environmental vulnerability of their sites of operation. Ceteris paribus, the higher the cost of being eligible faced by farmers operating in sensitive areas, the lower will be the environmental effectiveness of a cross-compliance measure.

In this respect, it is worth noting that a major potential problem with cross-compliance measures stems from the political difficulty of establishing mutual consistency between the original objectives pursued through the host programme and NPS pollution control requirements. For example, if the legislator intended to support low-income farmers, it is probable that when the host programme was designed, the beneficiaries were identified according to their economic status. However, to be effective (and efficient), cross-compliance measures addressing water pollution problems should be targeted according to those farm aspects that are environmentally relevant. It follows that cross-compliance provisions tied to environmental objectives may be difficult to reconcile with the host programme's original objectives, and their political viability may be undermined by opposition from targeted agents: this opposition will be all the stronger the more they feel themselves deprived of the right to benefit from a programme which other farmers with the same socio-economic status (e.g. acreage, or regional location) continue to benefit from (Dosi, 1994).

In May 1999, cross-compliance measures were introduced into CAP's instrument portfolio through Regulation 1259/1999.[10] According to Article 3 "Member States shall take the environmental measures they consider to be appropriate in view of the situation of the agricultural land used or the production concerned and which reflect the potential environmental effects. These measures *may* include:

- support in return for agri-environmental commitments,
- general mandatory environmental requirements,
- *specific environmental requirements constituting a condition for direct payments.*[11]

Article 4 specifies that:

"[Member States] may decide to reduce the amounts of payments which would ... be granted to farmers in respect of a given calendar year where:

- the labour force used on their holdings ... falls short of limits to be determined by the Member States, and/or
- the overall prosperity of their holdings during that calendar year, expressed in the form of standard gross margin corresponding to the average situation of either a given region or a smaller geographic entity, rises above limits to be decided by Member States, and or

- the total amounts of payments granted under support schemes in respect of a calendar year exceed limits to be decided by Member States."

Finally, Article 5 establishes the principle that "Member States shall apply the measures referred to in Articles 3 and 4 in such a way as to ensure *equal treatment between farmers* and to avoid market and competition distortions ".[12]

The principle of equal treatment, although politically appealing, does not, however, appear prima facie consistent with the need to take environmental heterogeneity into account when designing policy measures aimed at addressing NPS pollution problems. For instance, as already stressed, when environmentally oriented cross-compliance measures are designed, farm and farmland characteristics and location differentials (which affect on-site emissions and pollutant delivery rates) should be taken into account rather than farmers' socio-economic status. The achievement of other policy objectives, such as income support, should be pursued through other instruments, such as lump sum transfers.

Land retirement (set aside)

Land retirement, otherwise known as set aside, is one of the options available for reducing agricultural harmful impacts upon groundwaters.

As Ribaudo and Osrorn (1994) emphasize, one of the main justifications for a properly targeted land retirement programme is that the characteristics of NPS pollution and shortcomings in our ability to link on-field practices to environmental conditions make a practice-oriented approach very costly from an administrative standpoint. For instance, "while land retirement may be seen as unduly restrictive in that [social costs stemming from polluting activities] could be internalised and the land still remain in production, much lower administrative costs may justify its use when the merits of keeping the land in production are marginal" (Ribaudo and Orsorn, p. 85).

In both the United States (the Acreage Reduction Program, ARP) and in the European Community (EC Regulation 1094/1988),[13] set aside was initially introduced as a supply control policy instrument. However, with the 1985 U.S. Conservation Reserve Program (CRP) (and later on, with EC Regulation 2078/19920) the objectives of set aside were broadened, in that land retirement was seen also as an environmental policy instrument.

As far as the US experience is concerned, in the initial years of implementation, the CRP enrolled somewhat less acreage, and at somewhat higher cost, than originally planned (Rodgers *et al.*, 1990). One predictable source of ineffectiveness and inefficiency was the attempt to achieve too many objectives through a single instrument (set aside): soil conservation, supply control, and budget discipline. This and the fact that CRP competed with the more supply-oriented ARP, have led to inefficiency in both programmes. In an attempt to partly overcome these problems, Taff and Runge (1987) suggested a refinement in the eligibility criteria for both land set aside programmes: farmland which is highly productive but not environmentally sensitive should have been targeted through ARP, whilst farmland which is less productive but environmentally sensitive should have been targeted through CRP. Non-productive and non-environmentally sensitive farmland should have been made ineligible for both programmes. Other authors recommended adding additional targeting criteria, including an NPS index (Pearce, 1987), in order to take into account not only the gross erosion potential of farmland, but also geographical position in relation to potentially affected water bodies.

An additional criticism of the 1985 CRP was the lack of formal links between conservation provisions and the provisions of the federal Clean Water Quality Act (in particular, section 319, dealing with NPS pollution problems). Kuck *et al.* (1990), in particular, suggested that the Farm Bill's conservation provisions should be targeted to watersheds that the States, as required by the Clean Water Act, had identified as not achieving federal water quality standards because of residual NPS pollution from agricultural sources. The 1990 Reauthorization of the US Farm Bill partly accounted for these suggestions and criticisms by expanding the definition of eligible land to include areas subject not only to severe soil erosion, but also groundwater pollution.

The US experience (and something similar could be said for the EC experience) clearly shows that the effectiveness of the set aside programme, and, in particular programmes tied to NPS water pollution abatement, crucially depends on the regulator's ability to differentiate between eligible and non-eligible farmers based on farmland characteristics and location relative to sensitive water bodies.

Moreover, as with the cross-compliance provisions, the environmental effectiveness of set aside programmes depends on the degree of correlation between farmland productivity and farmland environmental sensitivity, as well as on the foreseen compensation for land retirement. In this respect, as Ribaudo and Osrorn (1994) have emphasized, to be cost effective, set aside programmes should mainly focus on "marginal cropland and/or cropland that discharges into particularly valuable water resources" (p. 85).

6.4. NPS pollution control: the direct regulatory approach

6.4.1. Ambient tax/subsidy policy schemes

Many authors have recommended the implementation of incentive mechanisms based on observable ambient pollution levels (groundwater pollutant concentrations) as an alternative to policies taking as a reference basis specific farming practices (input/output oriented policy instruments). The main rationale underlying this regulatory approach is that, similarly to individual effluent discharges, monitoring farming practices may be administratively difficult, or prohibitively expensive.

There is a key difference between policies which take specific farming practices as a reference basis (indirect regulatory approach) and policies based upon observable ambient pollution levels (direct regulatory approach). While "with policies that hinge only on firm-specific decisions ... , once the policy has been set, each firm does not have to consider its own pollution types or the types/actions of other firms since its own profits are independent of those types/action ... , with policies based on ambient pollution ... , each firm's profits will depend on ambient pollution, which is in turn a function not only of its own type/actions but also of the types/actions of other firms" (Tomasi *et al.*, 1994, p. 10).

The prototype of NPS pollution regulatory schemes based on observable ambient pollution levels is the tax/subsidy scheme proposed by Segerson (1988), where every farmer who is presumed to have contributed towards water quality impairment (or improvement) should be charged (or rewarded) according to the deviations between the measured and the desired ambient pollutant concentrations.

According to Segerson (1990), a main advantage of this approach is that it allows the desired water quality goal to be achieved in a cost-effective manner. Those farmers for whom changes in management practices would have little effect on water quality will not seek to alter their farming practices, whereas those farmers whose behaviour substantially affects water quality would be induced to take steps to reduce pollution. Moreover, those polluters for whom changes would be effective would have greatly flexibility to reduce pollution using techniques that are the least costly ones for their specific site characteristics. Finally, although it requires monitoring of water quality, a tax on ambient concentrations does not require either individual pollutant discharges or farming practices to be monitored.

As Bystrom and Bromley (1996) stress, the implementation of Segerson's environmental charge system is equivalent to forming associations within particular watersheds and making the group of farmers collectively responsible for water quality. For instance, if ambient pollution fees are levied, they are assessed against the collective as a group. This then forces the members of the group to monitor each other's behaviour, and to assess miscreants accordingly.

Despite the theoretical advantages, the direct regulatory approach suffers from several potential drawbacks. First, under this regulatory approach, taxes (or rewards) would be paid (or received) by every farmer, irrespective of his/her individual impact upon water quality. In other words, individual tax payments (rewards) will depend not only on her/his behaviour, but also on the behaviour of other polluters. This may raise legal or equity issues which could undermine the political viability of this incentive scheme. In particular, "ambient taxes can ... be difficult to accept for agents who have already lowered their emissions in the past and now have to pay charges for common emissions" (Millock *et al.*, 1997, p. 5). Second, this regulatory approach underlines the assumption that farmers possess adequate information about the nature and extent of their on-site emissions; moreover, farmers are assumed to be able to make correct evaluations about the ultimate impacts of their on-site emissions upon water quality. Both assumptions are somewhat questionable and, in any case, their feasibility should be assessed for each specific situation.

As far as the first assumption is concerned, farmers may not possess private (better) information regarding their on-site emissions, i.e. about the amount of pollutants originating on farmland which potentially may affect particular water bodies. This is especially true when potential pollutants begin as residuals of productive processes (e.g. nitrogen in excess of a crop's uptake). Consequently, they may be unable to properly identify the most effective (and efficient) management practices to reduce the generation of potential water pollutants.

As for the second assumption, farmers may be unable to predict with sufficient precision the cause and effect relationship between their management practices (their presumed on-site emissions) and the concentrations of pollutants observed in particular water bodies. In this respect, the direct regulatory approach appears to be potentially more appealing for relatively small watersheds, where few potential polluters operate, and for water bodies that do not rapidly flush out pollutants. It is however, much less suitable for large watersheds, where many farmers operate (sometimes together with other economic sectors), undertaking different activities (crop patterns, livestock and crop production, etc.) which may involve the generation of pollutants with complex transport paths.

6.4.2. *Investment in monitoring equipment*

As has already been emphasized, the major drawback of a regulatory scheme based on collective is that individual penalties (or rewards, if observed ambient concentrations do not exceed the desired water quality standards) will not only depend on individual behaviour, which cannot be monitored, but also on the behaviour of other farmers.

In an attempt to partly overcome this problem, Xepapadeas (1994) explored an alternative policy approach. He provides what to our knowledge is the first model that attempts to endogenize monitoring of individual emissions. In the literature on NPS pollution regulation, this is generally assumed to be either technically impossible, or prohibitively expensive. Xepapadeas (1994), however, considers the case where monitoring individual emissions is technically possible, so that information about individual effluents could, at least in principle, be improved by investing in monitoring equipment. Using a theoretical model, he explores the potential advantages (in terms of regulatory efficiency) of a policy scheme comprising of (a) an emission tax (based upon the observable part of individual emissions); (b) an ambient-tax *à la* Segerson; and (c) an investment policy (undertaken by the regulator) on monitoring equipment. Xepapadeas shows that under certain circumstances, the proposed policy package generate regulatory benefits. In particular, the increased observability of individual emissions lowers the ambient tax component, which is the component that is most likely to generate strong political opposition in the regulatory package.

The idea of endogenizing monitoring of individual emissions (rather than assuming than monitoring NPS discharge is either impossible or nearly always so) has been explored further by Millock *et al.* (1997). The main difference with Xepapadeas' model is that investment in monitoring equipment is not carried out by the regulator, but polluters themselves are induced, though appropriate incentives, to invest in monitoring in order to signal their true environmental performance.

Millock *et al.* (1997) show that the proposed monitoring incentive scheme aimed at inducing agents to exhibit their true characteristics, can be implemented even if the regulator is unable to monitor polluting input use on each farm, as long as the polluters can be identified. According to the proponents, the main advantage of the incentive scheme, which tends to transform part of the NPS pollution problem into a point source one, is that it would significantly reduce the regulator's information requirements.

6.5. Final remarks

Groundwater pollution has been identified as one of the major environmental threats faced by European countries. Even though other sectors may be the ultimate cause of pollution problems, the role of agriculture is not in doubt. Nonetheless, agricultural pollution is still far from being effectively addressed. This is partly attributable to the intrinsic difficulty of managing groundwater pollution from diffuse sources. But it is also attributable to the special status which has been granted to farmers who, generally speaking, have been exempted from credible mandatory regulations, and have not been confronted by economic incentives able to effectively influence their behavioural options.

Since agricultural impacts upon groundwater quality generally occur outside the borders of farms, and affect other individuals, policies which appeal to self-interest are useless. Similarly, appeal to social responsibility has rarely been an effective engine for substantial changes in polluters' behaviours. It follows that if, as asserted in various policy documents, NPS groundwater pollution abatement is one of the key European environmental objectives, major changes in policy styles and regulatory approaches towards the agricultural sector are required.

On the grounds on the available literature and the results of EC policies implemented during the last decade, the recommendable changes may be summarized as follows.

6.5.1. *Public awareness and NPS pollution control legitimation*

A major barrier to the implementation of effective policy measures is the lack of information about the nature, extent, and social costs of groundwater pollution from agriculturally diffuse sources. Public consciousness of agricultural impacts upon groundwater quality has to be raised to the same level as for pollution problems such as surface waters impairment due to effluent discharges from large and readily identifiable point sources. As long as those suffering from pollution are unaware of the short/long term costs of groundwater contamination, there will inevitably be a political bias in favour of polluters who oppose effective policy measures.

Institutional or legal barriers to the implementation of these measures often stem from what is probably the most definitive feature of groundwater pollution from agricultural diffuse sources, that is, the difficulty of identifying individual responsibilities. Since it is technically difficult or prohibitively expensive to acquire full information about individual discharges, alternative ways for identifying farmers' responsibilities, and for sanctioning, or discouraging environmentally harmful behaviours through appropriate incentives, have to be politically legitimated.

Policy options do exist for addressing the problem. One alternative would consist in granting regulatory legitimacy to available NPS biophysical predictive models. This would enable estimation of individual discharges so as to acquire a reference basis for mandatory regulations or economic instruments. An alternative approach envisaged by the economic literature would be to identify, with the aid of watershed-based models, the group of farmers who are presumed to be contributing to groundwater contamination, and then, through a bubble policy targeted to observable total pollutant concentrations, making the group collectively responsible for groundwater quality.

Although in principle appealing, both of the regulatory approaches may prove to be difficult to implement. In particular, as far as the former approach is concerned, in order to be effective and efficient, economic incentives aimed at penalizing or rewarding specific farming practices should be spatially differentiated so that the heterogeneity of environmental conditions can be taken into account. However, a properly targeted geographical differentiation of economic incentives may be administratively difficult or legally impossible, which explains why instruments such as input levies tend to lose much of their appeal when used in relation to NPS pollution problems, while spatial differentiation of mandatory regulations (e.g. compulsory implementation of codes of good agronomic practice in vulnerable zones) may prove to be easier to implement and more appealing.

As far as the bubble policy approach is concerned, its major potential drawback is that individual penalties (or rewards, if observed pollutant concentrations do not exceed a pre-identified threshold) would be paid (or received) by every farmer operating within a watershed, irrespective of his/her individual impact upon groundwater quality. This obviously raises legal or equity issues which could undermine the political viability of this incentive scheme. However, this drawback could be attenuated by combining a bubble policy with economic incentives based upon the available information on individual emissions, information which could be acquired through predictive models or through investments in monitoring devices.

6.5.2. *The polluter pays principle and the borderline between environmental services and the prevention of environmental damages*

Policies aimed at controlling pollution from agricultural sources have usually relied, and still largely rely upon what is often referred to as voluntarism, but which can probably be better described as a soft persuasion though subsidization approach. Besides being in contrast with the polluter pays ethics dominating other environmental policies, this approach has not brought about a significant and widespread reversal of pollution trends.

This ineffectiveness is at least partly attributable to the somewhat ambiguous distinction between farmers' environmental services and environmental damages, a distinction which should provide the legal basis for deciding whether or not farmers are eligible for compensation for farming adjustments.

In various EEC policy documents, environmental services (target levels, according to the Commission's terminology) are defined as the outcome of any environmentally friendly adjustment of farming which goes beyond the basic standards of environmental care (reference levels). Is this politically defined borderline between farmers' positive and negative environmental externalities equitable and consistent with the polluter pays principle? The answer obviously depends on which agricultural practices are interpreted as being part of the collection of farmers' rights.

Traditionally, the European Common Agricultural Policy has relied upon two implicit assumptions: (i) since agricultural production per se provides social benefits which exceed consumers' willingness to pay for agricultural products, farmers are entitled to be rewarded through subsidized output prices, and (ii) since farmers' endowments include the right to use their land as they want, any environmentally friendly adjustment of farming requires additional subsidization.

More recently, budgetary constraints, international trade disputes, and the increasingly popular idea that the environment belongs to society as a whole, have induced European policy-makers to slightly revise the traditional agricultural policy armoury by partly abandoning the system of guaranteed high prices for unlimited production, and by looking at alternatives for the rationale behind farmer subsidization.

The previously mentioned Commission distinction between target and reference levels may be seen as being one of the outcomes of this process, in that it reflects a partial reassignment of property rights between farmers and the rest of the society. According to this reassignment, farmers do not hold the right to use their land as they want, but hold the right to be rewarded for any adjustment of farming which goes beyond basic standards of environmental care.

As with any other politically constructed property rights systems, the system envisaged by the Commission is obviously questionable. What matters, however, is that to be credible and operative, this system requires a rigorous and unambiguous definition of the reference level in order to assess farmers' compliance with legal regulations, and to have a benchmark for identifying farmers' environmental services to be compensated by society.

However, in the EU in general, and particularly in those Member States which have not properly identified and credibly imposed basic standards of environmental care (e.g. failure to implement the Nitrate Directive), the in some ways intrinsically ambiguous distinction between farmers' negative and positive environmental externalities has reinforced the attitude among farmers that they should wait for compensation for any environmentally friendly adjustment of farming. Such a consolidated attitude is likely to make the implementation of cross-compliance measures, which, in principle, could potentially partly bridge the gap between farmers' subsidization and farmers' environmental performances, politically difficult.

6.5.3. *Integration between agricultural and environmental policies*

The need to include environmental protection into CAP has been acknowledged by European authorities, and has to some extent influenced recent reforms. The agri-environmental component of CAP, introduced through the 1992 MacSharry Reform, is destined to expand under the recently approved Agenda 2000 policy package. However, there is a clear need for better and more effective integration and coordination between agricultural policy, water resources management and environmental policy provisions.

Integrating groundwater protection objectives into CAP will, in practice, involve an ability to match agricultural policy more closely to environmental conditions by taking into account location differentials, and by tailoring policy provisions to the impact upon groundwater of alternative farming practices, rather than to the socio-economic status of farmers.

Rather than coming up with new Europe-wide specific measures, what is needed is a clearer European framework specifying the principle for a division of labour between CAP and environmental policy provisions, and between payments and regulation related to positive and negative externalities of agricultural production. Moreover, as Brouwer (Chapter 14, this volume) makes clear, in order to arrive at a concrete formulation of the groundwater conservation conditions that have to be fulfilled, a clear and unambiguous definition of the term 'good agriculture practice' is essential. Codes of good agronomic practice, properly defined by taking into account the heterogeneity of environmental conditions, could, inter alia, become benchmarks for deciding whether a farmer is or is not eligible for public support, e.g. in the context of cross-compliance recently included in the CAP instrument portfolio.

Notes

1. Following McCann and Easter (1999), transaction costs include: research, information gathering and analysis; enactment of enabling legislation including lobbying costs; design and implementation of the policy; support and administration of on-going programme; monitoring/detection; and prosecutio/inducement costs.
2. The potential impacts of agricultural policies on farmers' direct water use are more widely discussed later in the volume.
3. A special input that should be considered somewhat separately is water. Although water in itself is not a groundwater polluting input like nitrogen fertilizers and pesticides, its overuse may directly or indirectly contribute to aquifer depletion. As Giacomelli *et al.* (Chapter 3, this volume) point out, the potential contribution of irrigation to groundwater contamination is two-fold. First, aquifers are vulnerable to over-abstraction which may increase concentration of pollutants already introduced to the aquifer, or seawater intrusion in coastal aquifers. Second, deep-percolating water from irrigation contributes to aquifer enrichment by pollutants accumulated on farmland.
4. The *Future of Rural Society* (European Commission, 1988a) and *Environment and Agriculture* (European Commission, 1988b).
5. The roots can be very broadly identified in earlier measures such as voluntary set aside, experimental extensification and most significantly, in Article 19 of Council Regulation 797/85, which authorized Member Countries to introduce special national schemes in sensitive areas to subsidize environmentally friendly farming adjustments.
6. Total expenditure (1993–1998) for the implementation of Regulation 2078/92 is expected to reach nearly ECU 5.5 billion (European Commission, 1998).
7. To our knowledge, the first official EC document where cross-compliance was considered as a policy option to address agricultural related water pollution problems is the already mentioned Commission's proposal *An Action Programme for Integrated Groundwater Protection and Management*, where it is stated that: "all possibilities, including use of economic instruments in order to reduce use of manure and chemical fertilisers to the amount required for crop production and compatible with protection of the environment and fresh water quality should be explored .... The development of codes of good agricultural practice ... should be at the centre of action taken. ... As compliance with the codes in itself may not be sufficient to achieve the objectives in certain regions, measures of a further-going nature to ensure environmentally compatible production could be developed. Possibilities for using the principle of *cross-compliance* should be explored in this context" (European Commission, 1996; Action Line 3.2).
8. A cross-compliance measure was introduced at the Community level through Regulation 1765/1992, forming part of the McSharry reform package (see section 3.2.4, footnote 5). However, this cross-compliance measure was not targeted to environmental goals, but to the reduction of production of surplus crops.
9. Italics added by the authors.
10. The Regulation applies to payments granted directly to farmers under support schemes financed in full or in part by the "Guarantee" section of the EAGGF, except those provided for under EC Regulation 1257/1999.
11. Italics added by the authors.
12. Italics added by the authors.
13. The first EC set aside programme provided for subsidies to any farmer committed himself to retiring the whole or part of his land from crop production for at least 5 years. The main objective was to reduce production of surplus crops; despite certain claims on this subject, the

programme did not have (and was not designed to achieve) environmental goals. Through Regulation 1765/1992, forming part of the McSharry reform package, another set aside programme was introduced with the aim of reducing production of surplus crops; the main difference with respect to the 1988 programme lying in the fact that apart from certain categories of farmer, retirement of a certain amount of cropland was compulsory, or, more precisely, non-compliant farmers were made ineligible for price support programmes. In this respect, the 1992 land retirement provisions can be interpreted as the first cross-compliance measure adopted at the Community level, although the measure was not targeted to environmental goals. A voluntary environmentally oriented set aside programme was included in the agri-environmental programme established through Regulation 2078/1992.

References

Baldock, D., 1993, The concept of cross-compliance and its possible application within the common agricultural policy, London, Institute for European Environmental Policy, 3rd draft.

Baldock, D., Lowe, P., 1996, The development of European agri-environment policy, in: Whitby, M., ed., The European environment and CAP reform. Policies and prospects for conservation, Wallingford: CAB International, pp. 8–25.

Batie, S., Sappington, A.G., 1986, Cross compliance as a soil conservation strategy: a case study, American Journal of Agricultural Economics, 68, 880–885.

Börkley, P., Glachant, M., Lévêque, F., 1998, Voluntary approaches for environmental policy in OECD Countries, Paris: OECD.

Braden, J.B., Johnson, G.V., Bouzaher, A., Miltz, D., 1989, Optimal spatial management of agricultural pollution, American Journal of Agricultural Economics, 71, 404–413.

Brau, R., Carraro, C., 1999, Voluntary approaches, market structure and competition, Note di Lavoro 53.99, Milan, Fondazione ENI E. Mattei.

Bystrom, O., Bromley, D.W., Contracting for non-point-source pollution abatement, University of Wisconsin-Madison, Department of Agricultural and Applied Economics, Staff Paper 392.

Christensen, T., Rygnestad, H., 1999, Environmental cross compliance: issues for policy design, Paper presented at the SOM Conference, Frederiksdal (Denmark), 21–23 November, 1999.

Dosi, C., 1994, Water pollution from agricultural and urban diffuse sources: its nature, extent and control, Note di Lavoro 5.94, Milan: Fondazione ENI E. Mattei.

Dosi, C., Ferro, O., 1990, Italian agriculture, CAP and the environment, University of Minnesota, Department of Agricultural and Applied Economics, Staff Paper Series, 90–18.

Dosi, C., Moretto, M., 1993, NPS pollution, information asymmetry and the choice of time profile for environmental fees, in: Russell, C.S., Shogren, J.F., eds., Theory, modelling and experience in the management of nonpoint-source pollution, Norwell MA: Kluwer Academic Publishers, pp. 99–121.

Dosi, C., Tomasi, T., 1994, Introduction, in: Dosi, C., Tomasi, T., eds., Nonpoint source pollution regulation: issues and analysis, Dordrecht: Kluwer Academic Publishers, pp. ix–xvii.

European Commission, 1985, Perspective for the common agricultural policy, Com(85)333.

European Commission, 1988a, The future of rural society, COM(88)501.

European Commission, 1988b, Environment and agriculture, COM(88)338.

European Commission, 1996, Proposal for an action programme for integrated groundwater protection and management, COM(96)315 Final.

European Commission, 1997, The implementation of council directive 91/676/EEC concerning the protection of waters against pollution caused by nitrates from agricultural sources, Report COM(97)473, Brussels.

European Commission, 1998, State of application of regulation (EEC) No. 2078/92: evaluation of agri-environmental programmes, DGVI Commission Working Document (VI/7655/98).

European Commission, 1999a, Agriculture, environment, rural development: facts and figures – a challenge for agriculture', http://europa.eu.int/comm/dg06, July 1999.

European Commission, 1999b, Europe's Environment. What directions for the future?, COM(1999)543 Final, Brussels.

European Environment Agency, 1995, Europe's environment. The Dobris assessment, London: Earthscan Publications.

Goetz, R.,Zilberman, D., 1995, Mining the soil: agricultural production system on peatland, environmental and resource economics, 4, 1–20.

Goralic, S.M., Remson, I., Cottele, R.W., 1979, Management model of groundwater system with a transient pollutant source, Water Resources Research, 15, 1243–1249.

Griffin R.C., Bromley, D.W., 1982, Agricultural runoff as a nonpoint externality: a theoretical development, American Journal of Agricultural Economics, 64, 27–32.

Kuch, P.J., Jones, J.J., Wolcott, R., 1990, Conservation beyond the 1985 Food Security Act: Linking the Water Quality Act section with the 1990 Farm Bill, in: Napier, T.L., ed., Implementing the Conservation Title of the Food Security Act of 1985, U.S. Soil and Water Conservation Society, 334–342.

Lévêque, F., 1997, Voluntary approaches, Environmental Policy Research Briefs No. 1, produced within the framework of the EU Concerted Action on Market Based Instruments.

Lichtenberg, E., Zilberman, D., 1987, Regulation of marine contamination under environmental uncertainty: shellfish contamination in California, Marine Resource Economics, 4, 211–225.

Lichtenberg, E., Zilberman, D., Bogen, K.T., 1989, Regulating environmental health risk under uncertainty: groundwater contamination in California, Journal of Environmental Economics and Management, 17, 22–34.

McCann, L., Easter, W.K., 1999, Transaction costs of policies to reduce agricultural phosphorous pollution in the Minnesota river, Land Economics, 75, 402–414.

Millock, K., Sunding, D., Zilberman, D., 1997, Regulating pollution with induced adoption of monitoring technology, University of California, Berkeley, Department of Agricultural and Resource Economics, mimeo, November 1997.

Milon, J.W., 1987, Optimizing nonpoint source controls in water quality regulation, Water Resources Research, 23, 387–395.

OECD, 1991, Environmental policy: how to apply economic instruments, Paris: OECD.

OECD, 1997, Evaluating economic instruments for environmental policy, Paris: OECD.

Pierce, F.J., 1987, Complexity of the landscape, in: Halbach, D., Runge, C. F., Larson, W.E., eds., Making soil and water conservation work: scientific and policy perspectives, Aukeney, Iowa: Soil Conservation Society of America.

Ribaudo, M., Osrorn, C., 1994, Land retirement as a tool for reducing agricultural nonpoint source pollution, Land Economics, 70, 77–87.

Rodgers, C., Easter, K.W., Tomasi, T.G., 1990, The off-site economic benefits of soil conservation: a review and discussion of recent literature on the recreational demand for water quality improvement, University of Minnesota, Department of Agricultural and Applied Economics, Staff Paper Series (pp. 90–45).

Roy, A.D., 1952, Safety first and the holding of assets, Econometrica, 20, 431–449.

Segerson, K., 1988, Uncertainty and incentives for nonpoint pollution control, Journal of Environmental Economics and Management, 15, 87–98.

Segerson, K., 1990, Incentive policies for control of agricultural water pollution, in: Braden, J.B., Lovejoy, S.B., eds., Agriculture and water quality. International perspectives, Boulder: Lynne Rienner Publishers, 63–104.

Shah, F.A., Zilberman, D., Chakravorty, U., 1995, Technology adoption in the presence of an exhaustible resource: the case of groundwater extraction, American Journal of Agricultural Economics, 77, 291–299.

Shortle, J.S., Dunn, J.W., 1986, The relative efficiency of agricultural source water pollution control policies, American Journal of Agricultural Economics, 68, 668–677.

Taff, S., Runge, C.F., 1987, Supply control, conservation and budget restraint: conflicting instruments in the 1985 farm bill, in: Halbach, D., Runge, C.F., Larson, W.E., eds., Making soil and water conservation work: scientific and policy perspectives, Aukeney, Iowa: Soil Conservation Society of America.

Tomasi, T., Segerson, K., Braden, J., 1994, Issues in the design of incentive schemes for nonpoint source pollution, in: Dosi, C., Tomasi, T., eds., Nonpoint source pollution regulation: issues and analysis, Dordrecht: Kluwer Academic Publishers, pp. 1–38.

Xepapadeas, A., 1994, Controlling environmental externalities: observability and optimal policy rules, in: Dosi,C., Tomasi, T., eds., Nonpoint source pollution regulation: issues and analysis, Dordrecht: Kluwer Academic Publishers, pp. 67–87.

Zeitouni, N., 1991, Efficient management of groundwater quality, Natural Resource Modelling, 5, 293–320.

CHAPTER 7

Voluntary and Compulsory Measures to Implement a More Sustainable Agriculture in Water Catchment Areas

Ingo Heinz

7.1. Introduction

In most of the EU member countries, the environmental problems caused by agriculture have not yet been adequately solved, despite existing laws. More efforts are needed in the immediate future, especially with regard to the maintenance of drinking water supply. Although changes in the water quality can be observed in some water catchment areas, the concentrations of nitrates and pesticides in aquifers must be further reduced by changing agricultural practices. In regions where water resources are not yet substantially polluted, preventative measures have to be taken.

The number of regulations in individual EU member countries which are being influenced by regulations at the EU level is increasing. The most important of them have to do with the admissibility of pesticides (Directive 91/414/EEC) and to water pollution by nitrates (Directive 91/676/EEC). Other EU regulations have been established to promote more sustainable farming (EEC Regs. 2092/91 and 2078/92) or to alter the Common Agricultural Policy (EEC Regs. 1259/1999 and 1257/1999). As experience in relation to the Nitrate Directive of 1991 shows, the difficulties involved in applying these regulations in the individual member countries cannot be overlooked (Dosi and Zeitouni, Chapter 6, this volume). For example, this directive was not incorporated into German law until 5 years later, with the issuing of the *Düngeverordnung* (Fertilizer Regulations). Complete implementation of this regulation by the authorities in each of the German Federal States will take a considerable amount of time.

In the EU, the view that the enforcement of environmental standards must be accompanied by voluntary instruments such as co-operative agreements has become more and more popular. Contracts between farmers and water companies and/or water authorities, are increasing in significance, even though subsidy programmes to fund changes in agricultural practices on a voluntary

C. Dosi (ed.), Agricultural Use of Groundwater, 157–167.

basis have already existed for some time. These contracts appear to be decisive instruments in helping to move towards a more environmentally friendly agriculture. In particular, the implementation of agri-environmental policies becomes easier, or in certain cases only possible through such contracts.

In Germany especially, several instructive examples can be found of environmental politicians and lawmakers conceding the advantage of the voluntary approach compared with a rigid imposition of rules such as prohibitions or restrictions on farmers. They have acknowledged that by using this approach, a sustainable agriculture, especially with regard to water, can be obtained far more cost effectively than by using exclusively compulsory regulations.

In many other EU member countries (such as Italy, for example), voluntary agreements between farmers and water companies are rare or do not exist at all. The reasons for this, and the question of what conditions are needed to establish such agreements, are of particular importance. It is possible that the experience gained in Germany, the Netherlands or France could be used to create the prerequisites for a more widespread application of this relatively new agri-environmental instrument within the European Union.

However, the question arises as to what procedures are suitable for ascertaining the most appropriate measures and the most suitable sites and times for reaching the maximum possible cost-effectiveness. The point of departure of an on-going EU research project[1] is the assumption that voluntary agreements between farmers and water suppliers can contribute towards implementation of environmental targets in an environmentally effective and economically efficient way.

7.2. Different types of instruments to change farming practices in water catchment areas – German experiences

Farmers' behaviour in respect of the use of pesticides and fertilizers can, in principal, be influenced by the application of command and control measures (prescriptions and prohibitions), by economic instruments (taxes and subsidies) and by promoting self-regulation mechanisms (voluntary agreements and contracts) (Dosi and Zeitouni, Chapter 6, this volume; Brouwer *et al.*, 1999; Just and Heinz, 1999; Oskamp *et al.*, 1998). Regulations which determine rules for farm practices are currently the main policy instruments for controlling adverse impacts on the environment. However, measures to change farmers' behaviour which make use of economic incentives have become more and more important in recent times. This development is well documented, especially in the new EU agri-environmental policy (as mentioned above). By contrast, the instruments used to promote self-regulation (which implies the participation of farmers in the implementation of environmental targets), are unusual in most of the EU member countries. This approach is only widespread in the water

Table 7.1. Different types of co-operative agreements in the water sector

Participation	Water abstraction charge	Type
Farmers and water companies	No	A
Farmers, water companies and authorities	Yes	B

sector in Germany, and to a certain extent in the Netherlands. Similar approaches can also be found in France.

With respect to German experiences, there are different types of co-operative agreements between farmers and water companies which are based more or less on self-regulation (see Table 7.1).

Type A represents the case of "pure" self-regulation, i.e. water companies enter into negotiations with farmers in order to obligate them to take certain preventative measures. An extreme form of this is the purchase of agricultural land by water companies. Since this option is mostly not available for protecting the entire water catchment area, the water companies make contracts with farmers, who receive compensation payments and/or free advice. This type exists in the German states of Bavaria and North Rhine-Westfalia. Under the conditions of type B, the authorities play a predominant role. They determine the purposes for which the revenue from the water abstraction charge can be used. However, the water companies and farmers both participate in this decision process. In the German state of Lower Saxony, there are contracts between farmers and water companies as well as between farmers and authorities. In the first case, the water companies who compensate farmers are reimbursed by the authorities from the revenue from the water abstraction charge. A similar situation to type B can also be found in the German states of Baden-Württemberg and Hesse.

In contrast to type A and type B, which are based on voluntary, mostly contractual, agreements, in Germany water companies (or the authorities) are obliged to make compensation payments to farmers if water protection areas have been created. In these areas, the farmers are legally forced to meet higher requirements compared with farmers outside those areas. According to the German Federal Water Act (*Wasserhaushaltsgesetz*), the farmers can demand to be compensated in the event of income losses caused by lower yields and/or higher production costs. However, in some regions, the water protection areas are too small to completely prevent pollution of aquifers used for drinking water supply. In such situations, water companies often use the option of negotiating with farmers within the framework of either type A or type B. They also choose this course of action if the environmental rules, especially in water protection areas, are not specific enough.

A significant important example of this is how the EU Nitrate Directive was implemented in Germany through the fertilizer regulation (*Düngeverordnung*). According to many water companies, the upper nitrate limit of 170 kg/ha for

Table 7.2. Voluntary agreements with farmers in drinking water catchment areas and their application in Germany

Federal State	Area under agricultural use (1000 ha)	Total drinking water production (million m³/year)	Catchment of ground and spring water (%)	Number of water companies	Charging for water removal	Voluntary agreements with farmers*
Baden-Württemberg	1447	707	75	1267	Yes	3
Bavaria	3342	954	93	2585	No	150
Brandenburg	1345	141	87	137	Yes	0
Hesse	775	408	95	468	Yes	24
Mecklenburg-Vorpommern	1349	113	79	66	Yes	0
Lower Saxony	2675	571	87	346	Yes	107
North Rhine-Westfalia	1531	1420	38	594	No	110
Saarland	73	61	100	48	No	1
Saxony	896	306	43	127	Yes	1
Saxony-Anhalt	1167	131	53	89	Yes	0
Rhineland-Palatinate	718	249	92	245	No	7
Schleswig-Holstein	1034	220	100	560	Yes	0
Thuringia	801	191	61	118	No	1
Berlin, Bremen, Hamburg	24	333	100	5	Yes	2
Total	17 182	5810	73	6655		406

* Most of the figures refer to the number water catchment areas where co-operative agreements are in place (December, 1999), while a few figures refer to areas in which water companies are involved. Some figures are estimations.

manure spreading is too high. In their opinion, it is not necessary to impose a maximum upper limit higher than 80 kg/ha. Furthermore, there are no obligatory rules for the maximum storage capacity of livestock manure, so that the farmers' expenditure on such facilities has to be subsidized. In those German Federal States where water abstraction charges have been established, the water companies finance these subsidies indirectly, whereas in other Federal States (for example, in Bavaria or in North Rhine-Westfalia), they pay farmers directly for this. Consequently, for many water companies, voluntary agreements are the only way of ensuring that farmers fulfil higher requirements in water catchment areas.

Table 7.2 gives an overview of the occurrence of voluntary agreements applied to alter farming practices in water catchment areas in the German Federal States. As Table 7.2 shows, most co-operative agreements are located in Bavaria, Lower Saxony and in North Rhine-Westfalia, but many such agreements can also be found in Hesse and Rhineland-Palatinate. The proportion of groundwater and spring water in drinking water production is generally high in Germany, especially in those states where voluntary agreements between farmers and water companies have been established. One exception is North Rhine-Westfalia, where many survey waters are in agricultural areas. Because of the large water catchment areas, water protection zones as a compulsory instrument are far from being sufficient for water protection. In response to this, the government, together with associations of farmers and water suppliers, created a joint programme in 1989 that aims to promote co-operative agreements with farmers in all water catchment areas. These agreements are mainly based on agricultural advisory services paid for by the water companies. The Ministry of Environment, Planning and Agriculture of this state subsidizes certain measures taken by farmers who participate in the co-operative agreements. One of the most important principles of this joint programme is to protect and amend the water quality in agricultural areas without worsening the economic situation of the farmers.

In German states where agreements have been made, specific compensation payments to farmers are made, either from the revenue of the water abstraction charges or directly paid for by the water companies. The amounts spent on promoting more sustainable farming mostly depend on the concrete measures taken by the farmers and take into consideration the specific environmental pressures on the local waters.

A different situation exists in Baden-Württemberg where the farmers' behaviour is, first of all, determined by the rules which must be kept in water protection zones and where revenue from the water abstraction charges is predominantly used to compensate farmers at a flat rate. This means that the specific state of water pollution in individual localities is less decisive in farmer compensation. Furthermore, negotiations on suitable changes to farming methods on a voluntary basis (as the main feature of co-operative agreements between farmers and water companies) have less significance. Using such a compulsory instrument to influence farming methods might involve a disadvantageous and inefficient use of funds unless the farmers' practices can be controlled by a relatively exact monitoring system, continuously measuring data, especially on the contents of nitrates or pesticides in soil and water. Indeed, such a more or less complete monitoring system has recently been installed in Baden-Württemberg.

Nevertheless, the question remains as to whether co-operative agreements are preferable to the compulsory instrument combined with compensation. In voluntary agreements as a self-regulatory interactive process, mutual confidence, self-determination and locally specialized knowledge of the farmers, water suppliers and authorities play a predominant role. Furthermore, the readiness of the farmers to participate in agricultural advice programmes or to communicate with advisors is proving to be higher in a voluntary and autonomous atmosphere. In some German states (for example, Hesse and North Rhine-Westfalia), this insight has recently led to a change in policy on the part of the water authorities which has shelved enforcement of the rules issued in water protection zones in cases where voluntary agreements with farmers have been established. Farmers who do not join in with such agreements are forced

Table 7.3. Expenditure of a water supply company to reduce pollution caused by farmers (1998 estimated figures)

Payments	EUR/a	EUR/ha
Compensation for measures to improve the nitrogen balance by reduced use of fertilizers, to reduce the application of pesticides, substitution of mineral fertilizer by manure, intercropping, reducing livestock numbers, conversion to ecological farming	82 000	21.0
Subsidizing the purchase of equipment for a more sustainable pest control (e.g. maize hoe, curry-comb) and fertilizer application (e.g. drag hoses, semi-liquid manure reservoir)	77 000	19.7
Application of the nitrate min – method	102 000	26.2
Monitoring of pesticides	5000	1.3
Compulsory compensation for reduced use of pesticides	18 000	4.6
Management	72 000	18.5
Agricultural advisory services	31 000	7.9
Total	387 000	99.2

to periodically provide documentation to the authorities showing that they have changed their production methods according to the prescribed rules.

One example which shows the advantages of the co-operative approach in terms of groundwater protection and cost savings in drinking water production is that of a small water company named *Stadtwerke Viersen*, located in North Rhine-Westfalia. The groundwater removal amounts to 6 million m^3/year. There are four water protection zones with a total of 3900 ha of agriculturally used land. The payments of the water company to individual farmers and for other preventative measures are of special interest (see Table 7.3).

A comparison between expenditure on the preventative measures of 0.06 EUR/m^3 and the cost savings in water treatment of 0.28 EUR/m^3 for biological denitrification shows the economic advantage of this approach (Kooperation Landwirtschaft, 1998).

7.3. Co-operative agreements and the nitrate directive

The nitrate directive 91/676/EEC aims to establish a legal framework for more sustainable farming undertaken as a consequence of water pollution caused by intensive agriculture. The regulations which the Member States have to enforce provide for temporary bans on applying certain fertilizers, specify the carrying capacity of containers for manure and the maximum allowable amounts of fertilizer to be spread, require the monitoring of surface waters and groundwater, and require identification and designation of vulnerable zones and action programmes for such zones. The implementation of the directive by the member states took place later than originally scheduled, and to this point has only been partly implemented. The European Commission recently decided to submit an application to the European Court of Justice against many member states, including Germany.

The nitrate regulation (*Düngeverordnung*) of 1996 is aimed at implementing the directive in Germany. This regulation aims first and foremost at determining a general code for good agricultural practice. Although such a code has been in use since 1989, when it was prescribed in the German *Düngemittelgesetz* (the fertilizer law), it was insufficiently defined. There are now rules for the application of fertilizers according to plant nitrogen requirements, requirements established with the prevention or reduction of water pollution particularly in mind. The rules refer specifically to the spreading of manure: for example, they make it obligatory to first determine the total nitrogen, phosphate and potash contents, and in the case of semi-liquid manure, ammonium nitrate. These contents can be estimated by using approximate figures issued by agricultural authorities. Furthermore, the farmers have to maintain minimum distances from surface waters in order to avoid leaching of nitrates. These distances can be prescribed by the local authorities. Farmers are also obliged to analyse the nitrogen requirements for every plot, noting the different types of plants, soil features such as the nitrate, calcium and humus content, and farming practices which may influence the nutrient supply, such as the preceding crops, soil management and irrigation. Last but not least, farmers with more than 10 ha have to carry out periodic nutrient balances. Extensive farming with low livestock numbers is exempted from this obligation.

The way in which these rules are to be implemented by the individual farmers depends mostly on specific local conditions, including the type of soil, the slope of the ground, climatic variations, frequency of rainfall, irrigation systems applied, cultivation development, groundwater level, depth of soil, and features of surface waters, such as river stages. For water supply, the size of local water catchment areas and the distance to the water works is especially crucial. Consequently, it is necessary to have a code for good agricultural practice such as that formulated in the German *Düngeverordnung*, though by itself it is not adequate. Because of the manifold local differences in agricultural land, climatic conditions and water resource characteristics, a specialized code of good agricultural practice should be defined for every farming area. This is obviously not possible for individual farms. However, this can be carried out for agricultural areas with the same or similar conditions.

The German *Düngeverordnung* is mainly a compulsory agri-environmental instrument, but only a few of its rules can be enforced with command and control costs that are acceptable. Voluntary agreements with farmers which make use of various locally specialized codes of good agricultural practice can thus make essential contributions to the implementation of the *Düngeverordnung*. One possible approach is the use of subsidy programmes which are based on, for instance, the EU Regulations 2078/92 and 1257/1999, respectively. A further option is the use of co-operative agreements between farmers and water suppliers.

Although the *Düngeverordnung* plays a crucial role in clarifying codes of good agricultural practice, as an agri-environmental instrument, it nevertheless appears to have only limited effectiveness, especially with respect to the enforcement of water quality standards. This is mainly because of the rigorous requirements for ensuring drinking water quality. Co-operative agreements which take local conditions into account would therefore seem to be important in assisting in the conversion of farming to more environmentally friendly agricultural methods. In contrast to instruments such as uniform restrictions on applying fertilizers or charging for the use of fertilizers, the cost-effectiveness of measures to prevent water pollution as agreed to in co-operative agreements with farmers can be regarded as being especially high.

Furthermore, the *Düngeverordnung* is only a legal framework, because Germany applies the nitrate directive and the action programme throughout its territory. This could be the basic reason why the implementation of the nitrate directive is criticized by the EU Commission (i.e. the same high requirements cannot be prescribed in every region). The local authorities have to specify the rules to be met according to the specific conditions, especially in legal water protection zones.

These important factors explain why the authorities are increasingly relying on farmer awareness being promoted by agricultural advisory services rather enforcing compulsory measures. In North Rhine-Westfalia, as long as the farmers are willing to follow the obligations voluntarily agreed to with water suppliers, the authorities often do not systematically check whether they are complying with all of the *Düngeverordnung* rules or rules in relation to legal water protection zones. By signing the agreements, farmers in these cases have, in fact, declared that they will consider the nutrient requirements of soils, participate in $\text{nitrate}_{\text{min}}$ analysis programmes, measure the nutrient content of semi-liquid manure and prevent nitrate leaching into surface waters. Furthermore, they have agreed to observe the general code of good agricultural practice. In some cases, farmers have agreed not to spread semi-liquid manure from September 1st (i.e. 10 weeks earlier than laid down in the *Düngeverordnung*) and to give surplus manure to farmers with a low lifestock density. In return, they receive free advice and access to special subsidy programmes, such as those for financing semi-liquid manure reservoirs. In addition, the water companies subsidize the purchase of equipment such as drag hoses and manure reservoirs for more environmentally friendly manure management.

The obligations voluntarily agreed to are very often far stricter than the rules that have to be met in legal water protection zones. In some specific cases, organic farming practices have been put into practice.

The controls are predominantly run by water suppliers and the farmers themselves (self-controlling among the farmers). As a consequence of the obligation to carry out nutrient balances at farm level, agricultural advisors are able to show farmers how they can change their nutrient situation and convert to more economical fertilizer management. Only those farmers who are not willing to meet with the agreements, despite reminders and offers to advise them, will be punished by the authorities (Kühn, 1966). Similar regulations in which legal rules issued in water protection zones can be substituted (or complemented) by voluntarily agreed to obligations between farmers and water suppliers have been implemented in other German states (for example, Bavaria, Hesse and Lower Saxony),

7.4. Final remarks

While in Germany and in the Netherlands (but also in Austria and France) co-operative agreements between farmers, water suppliers and authorities are a suitable instrument for easing the implementation of agri-environmental targets, this option is rarely or never used in many other EU member countries. One aim of the EU15 research study mentioned above was to explore the reasons for this and to investigate the prerequisites for a more widespread application of this approach within the European Union (Heinz *et al.*, 2001). Such agreements are particularly scarce in the southern member states. This is surprising in view of the fact that water pollution caused by agriculture cannot be overlooked in those countries.

For example, no such agreements are to be found in Italy. In this country, there are arrangements which aim to use water for irrigation purposes as effectively as possible. However, drinking water suppliers (i.e. municipalities responsible for the water supply) do not participate in these Reclamation and Irrigation Consortia (Consorzi di Bonifica). The main reason for this is obviously the fact that there have been, at least until now, no severe conflicts between agriculture and public water suppliers over the use of the available water resources. In Spain, a completely different situation prevails. With respect to the growing concern about the water pollution caused by intensive farming and taking into account the increasing agri-environmental regulations at EU level, the question arises as to how voluntary agreements, as described above, might be a suitable solution to cope with these problems in these countries too.

In particular, the implementation of the nitrate directive demands that we look for further instruments which focus more on the different local circumstances to establish the most cost-effective measures. In some southern member countries, co-operative ventures which are now starting up could be expanded to incorporate a voluntary approach towards protecting drinking water against nitrate pollution. The reclamation and irrigation consortia which cover about 14 million ha in Italy already have a co-operative philosophy. There are also associations of different water users in Spain. The main purpose, however, is economical use of water for irrigation. But most of these associations (for example, the Irrigation Association of Castilla-La Mancha and the Irrigation Association of Castilla Leon) are also striving for more environmentally friendly agriculture.

In these countries, there is a tendency to strengthen voluntary agreements with farmers to solve environmental issues. These initiatives should be supported by the experience gained in Member States where such agreements are common practice (Heinz, 1999a, 1999b). In particular, the instruments already used by these governments to promote voluntary co-operative agreements should be assessed according to whether they could be transferred to the other member countries. These instruments include subsidizing the management costs of such agreements and compensating for specific farm income losses caused, for example, by not using fertilizers and pesticides near surface waters. A further political instrument could be to relate public subsidies (for the acquisition of drag hoses or semi-liquid manure reservoirs, for example) to participation in co-operative agreements. In certain circumstances, it might be useful to introduce a water abstraction charge in order to fund the changeover to more extensive farming, especially to preserve water resources which might be used in the future for the supply of drinking water.

The present efforts of the southern member states, including Portugal and Greece, in developing new regulations on agri-environmental and water policies, influenced additionally by the nitrate directive, should be viewed as an opportunity to establish the legal conditions favourable for creating voluntary agreements between farmers and water suppliers.

Note

1. European Commission-DGXII: "Co-operative agreements in agriculture as an instrument to improve the economic and ecological effectiveness of the European Union water policy" (Contract No: ENVA-CT98-0782).

References

Brouwer, F., Heinz, I., Zabel, T., 1999, The role of co-operative agreements in agriculture to achieve EU water policy targets, Paper presented at the 64th Seminar of the European Associations of Agricultural Economists, Berlin, October 27–29, 1999.

Heinz, I., 1999a, Contracts with water supply companies as an agri-environmental instrument in the European Union – theory and practice, Paper presented at the 64th Seminar of the European Associations of Agricultural Economists, Berlin, October 27–29. 1999.

Heinz, I., 1999b, How can co-operative agreements between farmers and water supply companies contribute to a more economically efficient compliance with environmental standards?, in Musy, A., Pereira, L.S., Fritsch, M., eds, Proceedings of the 2nd Inter-Regional Conference on Environment-Water Emerging Technologies for Sustainable Land Use and Water Management, Volume II, Ecole Polytechnique Federale de Lausanne, Switzerland.

Heinz, I., Brouwer, F., Andrews, K., Oltmer, K., Oldenbruch, R., Zabel, T., 2001, Co-operative agreements in agriculture as an instrument to improve the economic and ecological efficiency of the European Union water policy, Research Project for the EU Commission – DE Research, Contract No. ENV-CT98-0782.

Just, F. Heinz, I., 1999, Do 'soft' regulations matter?, in Brouwer, F., Lowe, P., eds, CAP Regimes and the European Countryside: Prospects for Integration Between Agricultural, Regional and Environmental Policies. Wallingford, CAB International. (forthcoming)

Kooperation Landwirtschaft/Wasserwirtschaft, 1998, Viersen: Stadtwerke Viersen GmbH and Landwirtschaftskammer Rheinland (publishers).

Kühn, H., 1996, Düngeverordnung und kooperativer Gewässerschutz, in Landwirtschaftskammer Westfalen-Lippe, Kreisstelle Coesfeld: Ein Bericht über die Ergebnisse der Beratung in 1996. Coesfeld: Kooperation Land- und Wasserwirtschaft im Einzugsgebiet der Stevertalsperre (publishers), 11–13.

Oskamp, A.J., Vijftigschild, R.A.N., Graveland, C., 1998, Additional EU policy instruments for plant protection products, Wageningen: Wageningen Pers.

CHAPTER 8

Negotiated Agreements Between Water Suppliers and Farmers in the Context of Changing Water Networks in Europe

Hans Th.A. Bressers, Stefan M.M. Kuks and Geerten J.I. Schrama

8.1. Introduction

Water problems make up a considerable part of the environmental problems with which our world struggles. Water quality is bad and worsening in the low-income countries and generally acceptable and slightly improving in the high-income countries (World Bank, 1992). In Europe, the improvement of drinking water quality has been the single most important cause of the increase of life expectancy (World Bank, 1992). Nevertheless, even in Europe water supply is threatened. In the European Union this is generally not a quantitative, but a qualitative problem. In particular the pollution of aquifers with nitrates and pesticides and the pollution of surface waters with nitrogen, phosphorus and various other substances gives cause for concern (European Commission, 1992). Meanwhile, the amount of water that is withdrawn for various uses is still increasing and the water has to be purified in ever more expensive ways. Nevertheless there remain uncertainties that hamper adequate action. These consist not only of uncertainties about the relationship between human dimensions and the deterioration of groundwater quality, but also of uncertainties about the ability to produce the desired changes in these human inputs. In this chapter we want to stress this aspect of the problem.

The possibilities for adequate action do not only depend on possibilities for direct government intervention. More and more such possibilities are even considered with some sense of criticism. Solutions are more sought in changing regimes and other institutions. We want to stress here that, besides their direct impact, interventions like financial incentives and general standards can also form an important institutional context in which changes in policy networks, regimes and more effective policy interventions can evolve. The European drinking water standards have been the driving force behind the recognition of the threats to the usefulness of groundwater aquifers (and surface waters) as sources for drinking water production in large parts of the European Union.

C. Dosi (ed.), Agricultural Use of Groundwater, 169–193.

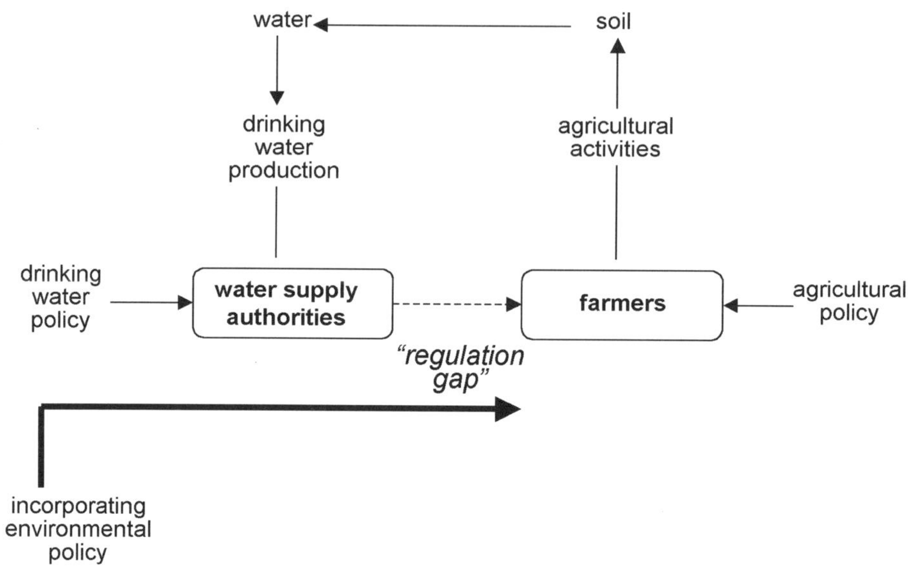

Figure 8.1. Regulation gap between water supply authorities and farmers.

Directly and indirectly, these standards have stimulated policy developments in various member states: without European standards, these developments would not have been obvious.

Such pressures are extremely important to open new windows of opportunities (Kingdon, 1984). Together with the pressure on public finance, the increased consciousness of the ecological challenge changed the policy networks of organizations in the water policy field in the 1980s in countries such as Great Britain, the Netherlands, Germany and the United States. They were all very different in many respects at both the beginning and the end of that period. Nevertheless they all became more open, more business-like and less dominated by an engineering orientation (Bressers and O'Toole, 1994). These changes, in their turn, have made also new innovative management strategies more feasible.

Drinking water standards, as generated by the EC Drinking Water Directive, force water supply authorities to do something. Whether they like it or not, they become part of the water policy network. However, such standards are only affecting the behaviour of water suppliers directly, and not the behaviour of those who are polluting water resources. Water suppliers need to fill in the regulation gap that exists between them and the farmers that are polluting (Figure 8.1).

In several European countries this leads to situations in which water supply authorities feel forced to negotiate with farmers on a reduction of agricultural pollution. To be sure that farmers do co-operate financial compensation is often paid (Kuks and Neelen, 1991). Water suppliers are then "buying the good behaviour of farmers" in water protection zones. As shown by Heinz (Chapter 7, this volume), such contracts appear to be decisive instruments in helping to move towards a more environmentally friendly agriculture. The implementation of agro-environmental policies becomes easier, or in certain cases even only possible through such contracts. A sustainable agriculture can be obtained far more cost effectively than by using exclusively compulsory regulations.

Despite the appeal of this approach, questions might be raised concerning the polluter pays principle. To prevent this violation of the principle, political interventions on the national and subnational levels are necessary. In this respect it is important to keep in mind the distinction between reference level (the minimum standard of environmental care) and target level (compliance beyond the reference level) as discussed in Chapter 6. In accordance to the EU borderline between polluter pays principle and polluter being paid, compensation to farmers should only apply in cases of over-compliance (where society asks farmers to provide an environmental service beyond the reference level). Such cases are cases where water companies want farmers to fulfil higher requirements (more effective adjustments in farming practices) in water catchment areas. An important desired effect of compensation for overcompliance is to achieve equal treatment of farmers and to avoid market and concurrence distortions.

A general institutional and policy context is the more important because fear for deterioration of mutual terms of competition forms a permanent threat to the legitimacy of strong environmental policies with the public and (in this case agri-) business. The danger exists that without such a context EU member states do not attune their environmental policies on the basis of an equal bottom line of environmental quality, but on the basis of equal costs for their economies. With an unequal degree of environmental deterioration and an unequal contribution to the threats to sustainability, unequal efforts for the environment seem justified however, and are to be viewed more and more as a normal part of the conditions of the place of business of a certain region.

Against this background the following subjects will be dealt with in this chapter. First we will deal with some changes in the policy networks of organizations involved in water supply and groundwater protection in the Netherlands, Germany, the UK (England and Wales were actually studied) and the United States that can be observed over the last two decades. Section 3 analyses more in detail the situation in the Netherlands. The Netherlands is an interesting case because it combines high aspirations on environmental policy and a flexible attitude towards policy innovations at a general level with an actual environmental situation regarding the agricultural pollution of the soil that is among the worst in Europe. In Section 8.4 we will report on an EU sponsored study (EV5V-CT94-0368), that investigated the possibilities for water supply authorities in Europe to prevent water pollution from agricultural sources. This research project included again case studies on the Netherlands, Germany and Great Britain, as well as by a United States case study. Section 5 summarizes the main conclusions and policy recommendations.

8.2. Changing water networks in Europe

The water policy networks of the Netherlands, Germany, Great Britain and the United States show both remarkable similarities and intriguing differences. In this section we attempt to analyse how they arrived at this. Which trends can be observed in these countries over the last two decades and what dynamics of change were associated with them? An overall conclusion is that fairly similar trends evolved in the four countries with established water policy networks, patterns basically emanating from more or less the same challenges. Diverse initial situations and a number of rather stable factors that influence both these circumstances modify these trends and the ways in which new challenges have been incorporated into network operations.

8.2.1. *Developments*

In the four countries water policy networks generally have become more open, more business-like, in the sense that certain organizations like water authorities try to behave more like businesses, and less dominated by an engineering orientation. These trends seem to be related to each other.

The British case shows, for instance, that the institutionalization of the more business-like identity of the water sector, culminating in large-scale privatization, brought the sector into the midst of the public debate, increased the range of organizations involved, and made the sector more vulnerable to external influences than in any time in its previous history. The environmentalists, especially, gained influence, simply by being out there. Dealing with these and other interests, forces the network actors to give more attention to social interactions and processes in their strategic considerations, and less to their older, predominantly engineering orientation.

The growing openness of the sector is apparent even in the US case, which was already permeable to begin with. New professions have been included. Even on the sub-national level the relationships among the actors involved are sufficiently loose so that even the concept of issue networks suggests more coherence than can be observed. In this case the entrance of new professions into the patterns is also related to more business-like water management. The shift is reflected, in other ways, by the emergence of privatized services. These new professions also bring new orientations, for instance an emphasis on economic efficiency.

In the Dutch case both the water supply companies and the water boards, central actors in two rather separate sub-networks, have declined sharply in number. The pressure for viable and efficient entities, which could be managed as modern businesses, has induced this shift. While water boards have always been more or less autonomous, indeed even guaranteed independence by the national constitution, the water supply companies have become significantly more autonomous. Nevertheless both water supply companies and water boards face an increasing need to respond to various kinds of external demands. This trend forces them to be more communicative, thus acting more on the basis of social interaction orientation and less from an engineering perspective. To achieve their new goals they have to negotiate with actors from outside the water world, actors that to some degree become thereby part of it. Thus far this form of openness has not really endangered the coherence of the network core. In the case of the Dutch water supply network signs are actually evident that the companies have come to accept more co-ordination through their association, as they face another well-organized community: agriculture.

In Germany the water networks, separate for various parts of the water cycle and in various Länder, have remained relatively stable. Although here too environmentalists have participated in discussions on a regular basis, it remains to be seen to what extent environmental interests manage to establish themselves as members of water policy networks. Furthermore, all over Germany new public, and sometimes private, organizations with large discretion and independent management are apportioned a share of water management tasks. Water supply utilities are dependent on the success of ground and surface water protection. The tension between groundwater protection interests and agriculture has not had as many consequences for the decline in dominance of engineering as in the Netherlands because in Germany water management already had often been included in environmental management agencies.

The European Union case also exhibits the phenomenon of expansion in types of participants involved in water policy, even though these constellations are still in the process of formation. The developing links with potentially large numbers of diverse actors are exemplified in the broad list of the DG XI General Consultative Forum.

8.2.2. *Change agents*

Apart from their mutual interaction, the developments described above can be related to a number of factors. Among these are the historic and geographical settings in which network developments have occurred, the impact of federalism in the US and Germany, German and European unification, and the notable lack of political salience of this sector in earlier years. Many of these, however, are more or less stable features of the national context and cannot be invoked to explain network changes (though they may affect the fashion in which these changes occur). Their influence is felt more on the ex ante situation and as intervening variables between the real causal forces or change agents, and network developments.

What then are these change agents? Two complex factors seem to lie at the heart of many of the observable lines of influence over network evolution: the welfare state crisis in the 1980s, especially in its public finance pressures, and the environmental challenge. Both have empirical and ideological dimensions, which vary across the countries in specific detail and intensity but have presented substantial meta-challenges in North America and Northwestern Europe during the 1980s and 1990s. Together they have had an unmistakable impact on the water sector: the time for pumping and billing is over.

The welfare state crisis became manifest after the first and especially the second oil crisis, contributing to stagflation. The neo-conservative response of the Thatcher and Reagan administrations set the tone for the direction of policy response: more market, less government. In other countries like the Netherlands and Germany, the ideological aspect of these policies was weaker. But less explicit, common sense notions stemming from renewed confidence in the capabilities of private business diffused in these countries as well. This shift resulted in the reorganization of water tasks, and in some cases a reduction in financial support and an increase in expectations for more business-like management of public tasks, even to the extent of privatizing the tasks in certain instances.

The environmental crisis emerged in two waves of public attention, one in the late 1960s and early 1970s, the other in the late 1980s and early 1990s. Although both these waves were triggered by epistemic communities warning on the basis of scientific studies of imminent environmental decline, each also carried an ideological dimension. After each upsurge, public attention declined but stabilized at a higher level than before, in the process creating both new governmental and private organizations and bases of power for them. General environmental awareness rose, of course and, in addition, some specific signs of environmental threats in the water sector increased the salience of the issues and added pressure for policy response. For instance, surface water pollution in such places as the Rhine had killed aquatic life and had often prevented the use of these waters for water supply purposes. Groundwater sources, as well, had showed increasing amounts of pollution which threatened to make them useless for drinking water production. The water sector inevitably had to deal with these problems, and the sectoral responses themselves caused new pressures. More generally, the tendency has been for other actors in other sectors, such as agriculture, to be drawn into water policy issues and thus render the networked context even more diffuse.

In short, the public finance challenge prompted, in particular, institutional and cultural developments promoting more business-like management; the environmental and scientific challenges encouraged a social interaction orientation; and both factors directly and indirectly forced more openness in the water sectors in several countries.

8.2.3. *Variety*

Similar developments stimulated by similar factors may present a misleading picture of uniformity, one that is valid on only a very general level. Closer inspection reveals considerable variety. The fashion in which the two fundamental factors just discussed have influenced developments in the four countries is influenced by features characteristic of each country.

For example, in the Netherlands and Germany many of the benefits possible from privatization were attained instead by greater autonomy and shifts in management culture among organizations in the water sector. The same process was happening in Great Britain, but other factors pushed the institutionalization of autonomy further, to full privatization, without there being a deliberate government policy toward this end. The British government became trapped in its own ideology. Though a four case study design is unlikely to demonstrate definitively the impact of these kinds of influences in a comparative fashion, some expected influences can be plausibly related to the observed differences in developments across the cases.

The highly legalistic culture of German policy making, for instance, seems to have affected the nature of the actors involved (e.g. law experts) and the kind of relationships among actors within the network (relatively inflexible). In fact, the changes in openness and orientation within the German water sector seem somewhat more minor than in the other countries.

So it is important to consider the significant differences in the ex ante situations across these four countries. These initial conditions are related to similar kinds of variables, factors that both help to determine the ex ante circumstances and also modify more directly the relationship between the main change agents reviewed earlier and the network developments of interest. Factors like natural or geophysical differences, plus scale and degree of federalism, shape antecedent circumstances and newer developments alike.

Not all such factors with a general impact create variety. Two that do not can be mentioned for purposes of the discussion below. The first of these is water itself, which actually flows through its own natural cycle, thereby suggesting and stimulating an undercurrent of interest in more co-ordination in the sector. Left to their own devices, the technical specialists of the sector in each of these countries would integrate their efforts through the cycle's phases. Thus factors promoting fragmentation never have an easy or permanent victory. Second, to a significant extent, the water sector operates specific technologies of its own. These encourage the sense that water engineers as a professional group are both distinct from other professional groups and able to harmonize water management throughout the developed world. The existence of such a tightly knit and technically specialized professional community has made it possible for the sector to be regarded as an apolitical, management-focused cluster, during certain periods at least; although the evidence from the EU is that newly forming arrays in the current period are likely to display more diverse characteristics even in the early stages.

One of the factors that does cause variation across the settings is the degree of natural diversity present, especially in combination with each country's history. The United States is home to almost every imaginable water circumstance. National uniformity cannot be a practicable aim under such conditions. The debate on subsidiarity within the EU echoes this theme, of course, with talk of repatriation of some water laws and of greater flexibility for the member states.

Another factor also creates some variation between the United States and the three European countries: the evolving European Union. To some extent it might be considered a change agent itself, because the impact, even interference of European regulation with member states' water policies is on the rise. The EU influence on the development of the water networks in these countries, however, seems more indirect and variegated. In Germany the reactions of the Länder to the European regulation differ from that of the federal government. This range of response has the effect of rendering the European dimension visible in German water politics. In Great Britain institutional developments have made the European regulatory issue much more visible. Here, as well, the European dimension became highly visible as a consequence. In the Netherlands, on the contrary, European regulation is completely included without much debate into national policies and standards. Its visibility as an exogenous factor is therefore low during everyday elaboration into regional policies and implementation, and its influence on the structure of the network not distinguishable from that of internal policy developments.

The initial situations in the various countries also seem to have had some influence on the ways in which the prime change agents have influenced network developments. Network stability depends not only on the degree to which the structures adapt to new challenges, but also on the extent to which these emergent issues become integrated into the professional expertise and attitudes of those in the network. In all countries examined here, as well as in the EU, environmental considerations have become more prominent than ever. The manner in which they have, however, has varied. In some nations' networks, environmental values were incorporated into the existing organizations; in others new organizations, with environmental issues as their prime focus, were added to the network. Access for new environmental organizations was not easy in any of the instances. In England and Wales their influence seems to stem primarily from others' taking them into account as a relevant outside force. In the Netherlands and Germany, the two most intact policy communities of the sample, environmental values seem to have been internalized to a larger extent by existing network organizations that have identified themselves with these new tasks. In the US both patterns are visible across the differentiated network waterscape, although the principal method has been the inclusion of new actors representing heretofore excluded or under-represented interests. The suggestion here is that the initial degree of network coherence reproduced itself in the ways in which environmental perspectives have been incorporated.

Even in the Dutch case more openness can be observed, but this shift has an impact mostly as a means to include new fields of expertise. The most typical pattern is that others are invited to participate in forums of the more established network participants. With considerable less frequency do the newer participants come alongside the older network, thus challenging it from the outside. This pattern is plausibly related to the fact that the Dutch networks are far less fragmented than those in the US to begin with. In the next section we will treat the Dutch network in more detail.

8.3. The Dutch water supply sector growing into a policy community

In the Netherlands the water sector has always been divided into two rather separate networks: on surface water management and on drinking water supply (Bressers *et al.*, 1995). Of the two, the water supply sector has been less coherent than the surface water management sector. This section will describe the evolution of the Dutch water supply sector into more coherence, facing-up the challenge of confrontation with other societal interests like agriculture and responding with extensive consultations with these interest groups.

8.3.1. *The evolution of the water supply sector*

There are many ways in which the coherence of networks can be characterized (Van Waarden, 1992; Kenis and Schnieder, 1991; Jordan and Schubert, 1992). Before them authors often stressed the intensity of network relationships (Dietz and Ryecroft, 1987) or, on the other hand, the fragmentation of many policy areas (Heclo, 1978; Kingdon, 1984). Here we will characterize the dimension of integration versus fragmentation by two variables, a structural variable (inter-relatedness) and a cognitive/affective variable (mutual commitment) (Bressers and Kuks, 1992).

The structural variable of inter-relatedness is the intensity and stability of mutual interaction. This interaction may consist of written and verbal communication, but also of exchange of personnel and the existence of formalized meeting groups and active intermediaries, which aim at an improvement of the contacts within the network. The cognitive/affective dimension of integration versus fragmentation can be termed commitment: the extent to which individuals, groups and organizations within the network sympathize with each others main objectives, as far as relevant to the policy area, and to which their cognitive maps of the policy area correspond. Both variables are of course among others dependent on how the network is defined: which interests are regarded as included?

Environmental interests have only recently gained attention within Dutch groundwater management (Kuks, 1988). Initially, groundwater was managed only for reasons of supplying drinking water and for related reasons of health care. For that purpose the Ministry for Health Care introduced the Water Supply Act in 1957, not only to make demands on the quality of delivered drinking water, but also to institutionalize the organization of the drinking water sector. This act formalized an already existing practice in which provinces can allow or forbid the establishment of new or extension of existing water supply companies.

Until the turn of the nineteenth century, most of the water supply companies were local and private initiatives. In the early twentieth century interlocal companies were established, often with the participation of municipalities. Municipalities and private persons were not willing to co-operate in all cases, however. To guarantee the efficiency of water works all over the country, several provinces developed their own initiative regulations for water supply, a practice later underpinned by the Water Supply Act.

Since 1975, most of the provinces have made plans for a further concentration of water supply companies. The number of companies was reduced from 102 in 1980 to 20 in 2000. In the future, a further reduction to maybe five is expected. The provinces and the drinking water sector itself, as represented by the VEWIN (Union of Water Supply Companies in the Netherlands) do agree that the structure of the drinking water sector should fit the demands for securing clean water supplies in the future. They think that water supply companies can only maintain their tasks if they have a strong organization, which implies sufficient technological know how and financial capacity.

The supply companies that use surface water for the production of drinking water have more problems with guaranteeing a good quality than those who use groundwater do. The latter consider themselves to be relatively invulnerable, and that is why there is a lot of resistance among them against reorganization plans. Their arguments are that they have never had problems with the supply of water, and that they always have had a good quality, that their charges are reasonable and that their customers are still satisfied (Van der Knaap, 1987). The smaller companies, mostly without the participation of provincial authorities, particularly try to maintain their autonomy, but they face the burden of proof to demonstrate that they can still operate in an efficient way. The VEWIN is very cautious in taking a stand (VEWIN, 1989): they support the idea of developing more professionality and efficiency in the drinking water sector. VEWIN tries to avoid a confrontation with the smaller companies, since it wants to be an organization that represents the entire drinking water sector.

To summarize, until the 1950s the water supply sector might be qualified as a rather fragmented network. Although the companies shared a common purpose, they acted separately. Water supply companies arose as local initiatives and for a long time they wanted to maintain their autonomy. Even today, the smaller ones are still fighting against provinces to revise reorganization plans. After 1957, the sector became more integrated because of two developments. First, reorganizations and merges between companies caused scale enlargements in the sector and an increasing inter-relatedness. Second, the inter-relatedness and mutual commitment within the sector increased because of the strengthening of organizations that were developed to support the collectivity of companies, such as VEWIN (Union of Water Supply Companies in the Netherlands) and the KIWA (a research institute for technological innovations in the water supply sector). The need for increasing efficiency in the sector encouraged a strong policy community with a strong technocratic approach to problems that the sector was encountering. Thus, these institutional changes were generated from inside the policy community, based upon a common perception of the way forward.

8.3.2. *The water supply sector encountering the agricultural sector in an issue network*

A serious threat to drinking water supply is the presence of nitrates in groundwater. More than 50% of Dutch withdrawal locations (especially in the east and the south) are likely to become unsuitable as a drinking water source in the near future. They are situated in areas with a sandy soil and many intensive cattle breeding farms (which tend to spread more manure on the soil than is necessary). Although there were indications that excess manure was being produced in some regions in the Netherlands as early as the mid-1960s, it took until the 1980s, because of a competence struggle between the Ministry for Environmental Protection and the Ministry for Agriculture, before the government seriously began to deal with the manure problem. The Soil Protection Act, which came into effect in 1987, is the first piece of Dutch legislation that aims for integrated protection of the soil and the underground water. It is primary aimed at preventing excessive manuring, by intervening in the manure spreading on agricultural land. The Act contains a number of standards to fix the amount of manure that is allowed to be spread. These standards apply nationally.

The Soil Protection Act also provides additional protection for areas in which groundwater needs to be withdrawn for the supply of drinking water. Provinces are authorized to establish so-called ground water protection areas and to enforce in more restrictive manuring standards in these areas. The Act further provides that farmers within the protected area should be financially compensated for the losses (disposal costs for the surplus of manure) they suffer as compared to farmers outside the protected area. Requests by farmers for compensation will be dealt with by the provinces, which can collect funds for this by means of a charge paid by those who abstract ground water (mainly the water supply companies). The levy is related to the amount of water abstracted. Water supply companies can charge the consumers of drinking water for this levy. In fact, the consumers of drinking water pay for the production of a collective good (according to the profit principle).

The reason for the compensation provision was to prevent protests of farmers in groundwater protection areas who encounter more restrictions than farmers outside those areas. Legislators feared that the more restrictive standards in groundwater protection areas couldn't be enforced without compensation. The political parties on the left opened the discussion on a motion to reject the provision because it implicitly would admit a right on pollution. The provision was felt to contradict the polluter pays principle. In the event, the equality principle took precedence. The consumers of drinking water seem to have had no voice in this political debate. They pick up the costs of pollution caused by farmers, as they do in other countries like England and Germany.

The drinking water sector thus encounters a very strongly organized opponent from outside the sector, as far as the farmers are concerned. Although the agricultural sector only forms 5% of the active working population in the Netherlands, it has considerable influence in Dutch politics. Since 1954 the agricultural sector is nationally organized through the Agricultural Board. The agricultural lobby is very effective, which is, for example, reflected by the fact that it usually is consulted in a very early stage of policy making. The Agricultural Board also has Regional Boards in each province that consult with the regional or local authorities, particularly about planning and land use. The regional boards play an important role in determining the manure policy for groundwater protection areas. They negotiate with the provincial authorities and with the water companies involved concerning the manure restrictions applicable in these areas, on the disposal of the resulting manure surpluses and how farmers are compensated for the losses they incur.

While the agricultural sector is strongly developed both at the national level and the regional and local level, the drinking water sector for a long time was not. Traditionally, water supply companies are organizations that are proud of their autonomy. VEWIN always played a modest political role. At best it was only active in emergencies. Normally, VEWIN mainly has a 'service' function with respect to the drinking water sector. At the end of the 1980s, VEWIN became more politically involved and responsive, although the political discussion on the Soil Protection Act had already been concluded. In addition, because of a presidential change (the presidency of VEWIN was taken over by the governor of the province of Zuid-Holland), VEWIN succeeded in moving the negotiations concerning the compensation in groundwater protection areas from the regional level (where one negotiator from the Agricultural Board representing the whole the country negotiated with each water supply company separately) to the national level. Since then, the framework for regional negotiations is the state level.

To summarize, external interests and pressure on the water supply sector in some ways strengthened the sector as a policy community. It became a more tightly organized policy community, especially in terms of an increasing commitment between the members of the community. In the meantime, however, the issue of agricultural pollution of groundwater confronted the water supply community with another strong policy community, the agricultural sector. This sector is very experienced and has a long tradition in lobbying and negotiating strategies. The agricultural sector as a network is not only characterized by a strong commitment, but also by a strong inter-relatedness. This strong inter-relatedness forced the water supply sector to participate in negotiations on a more aggregated (regional and national) level. In fact, it strengthened the inter-relatedness within the water supply sector.

8.3.3. *Consultation between water supply companies and farmers' organizations as the most promising strategy*

The discussion of the compensation provision shows that the regulatory strength (control capacity) of the authorities is very limited in the case of groundwater quality management. Although they try to regulate by means of ordinances and prohibitions, the enforcement of these rules is difficult. Therefore, a system has been chosen in which private organizations (water supply companies) have to participate in the enforcement of the rules.

The drinking water sector is beginning to define its role, however. Mr. Th. Martijn, director of VEWIN, stated: "This is a considerable change for organizations which traditionally are engaged in pumping and presenting the bill. However, if you want to create groundwater protection areas, then others may have fewer opportunities to use the soil in these areas. If that is the case, groundwater protection can only be realized by offering compensation to them. This is not a new phenomenon: the drinking water sector already contributes a third (about 20 million guilders) to the costs of the Rhine Salt Treaty. In the third National Water Management Directive this is called 'paying for quality'." (Jehae and Van Soest, 1990).

The compensation provision formally implies that farmers can claim their losses against the province, which in its turn may charge the water supply companies that withdraw groundwater in that area. Water supply companies prefer to settle the matter in a friendly atmosphere. They also want to control the disposal costs of manure surpluses and to avoid unnecessarily high bills. For both reasons, several water supply companies have started to meddle with the disposal of manure surpluses, which in itself is a very unusual task for a water supply company. An increasing number of companies is trying to buy out farmers in areas that are most vulnerable.

Although methods exist for purification of groundwater in the case of pollution with nitrates, the drinking water sector strongly opposes this option as long-term solution. The Director of VEWIN explains: "If the water supply companies started with complete purification tomorrow, the polluters could no longer be forced to change their behaviour. We want to use the drinking water, and with that the consumer, as a crowbar for improving the environment" (Velema *et al.*, 1989).

The strategy of consultation, chosen by the water supply companies to deal with agricultural pollution of groundwater, had already been tested by the companies needing to use surface water for their drinking water production. Those companies are mainly located in the western part of the Netherlands.

They have huge problems with maintaining a good water quality: a lot of the pollution in these rivers stems from foreign industries, which means that they are dealing with extra-territorial actors. The director of VEWIN indicated that VEWIN only reluctantly develops new techniques for analysing water for the purpose of tracing polluters. However, the data often can be used to exert pressure on polluting industries. For example, the City of Rotterdam has chosen to consult with polluting industries, even when they are abroad, rather than taking judicial action. Mr. F. Feith, of the City of Rotterdam, stated: “We try to handle the collected data very carefully, just because we want to get in conference with the discharging industries. Negative publicity will be applied only when the polluter is really unwilling. However, threatening publicity has proved to be a very strong instrument.” (Jehae and Van Soest, 1990).

This demonstrates that the drinking water sector realizes that it has its own role in water management, and that it can often reach further than any other authority. This seems to be true for the case of point source pollution, but is this also true for non-point source pollution? Consultation with polluters was regarded to be the only solution to the problem, since the national and provincial authorities were unable to guarantee strict enforcement of regulations.

Finally, we may formulate some conclusions on the way in which the water supply community dealt with external threats. On the one hand, water supply companies reacted in a very technocratic way by searching for technological innovations to satisfy the demand for drinking water of an acceptable quality. On the other hand, they tried to react against the Soil Protection Act, which ignored the polluter pays principle. They did not succeed in their opposition, although they were strongly supported by environmental groups. The water supply sector and the environmental groups have in common that they support the polluter pays principle. In this sense, the strong relationship between the water supply sector and environmental groups can be conceived as the existence of a broader policy community. However, the inter-relatedness in this community is weak: no strong or intensive interactions exist between both sets of actors. At the end, the polluter pays principle was not applied due to a successful lobby of the agricultural sector and due to the lack of political organization and influence of the drinking water consumers. It appeared to be the most feasible political outcome to charge the consumers with the costs of pollution prevention.

Another indication for the existence of interests that are common to the water supply sector and environmental groups is that they both stress the importance of strict rule enforcement. However, water supply companies do realize that it is very difficult to control the spreading of manure. The control capacity of the regulatory agencies (provinces) is limited with respect to this. That is why the water supply sector does expect better results through direct negotiations with farmers' organizations. After the settlement of the compensation provision in the Soil Protection Act, consultation with target groups was left as the most promising strategy in the issue network in which the water supply sector and the agricultural sector both participate. But that goes for the Netherlands. What would be the situation if considered in a broader European perspective? For that we returned to a comparative analysis. This was enabled by a research grant (EV5V-CT94-0368) of the SEER II European environmental research programme.

8.4. Water supply authorities preventing water pollution from agricultural sources

This section reports on the study 'Water supply authorities in Europe preventing agricultural water pollution' (for a more extensive report see Schrama, 1998). The project involved a comparative study of Germany, the Netherlands, and Great Britain, with special attention to the EU context. The project made it possible to learn from the experiences of water supply authorities in these countries in preventing agricultural water pollution. However, the problems that water supply authorities have with preventing agricultural water pollution are not specific for these countries. In most European countries water suppliers are faced with the same sort of problems. Therefore many results of this research project might be of relevance for other member states and the EU as a whole.

In this research project water supply authorities were defined as the organizations which are responsible for the supply of drinking water. Confronted with increasing scarcity of unpolluted resources, and high standards based on EU Directives, these organizations feel themselves forced to negotiate with polluters. In the case of agricultural pollution, financial compensations are often paid to farmers in exchange for their co-operation. Rather than denying or denouncing this phenomenon, the aim of this project was to investigate by which means environmental policy makers can strengthen the control capacity of water suppliers in their relation to farmers in order to affect the behaviour of farmers indirectly.

The policy network approach was used again as the theoretical framework. In the national case studies the water supply and agricultural sectors and their interlinkages and dependencies were analysed in terms of policy communities and issue networks. The concept of control capacity, the dependent variable in the study, was analysed from the perspective of the distribution of critical resources. As a background factor some characteristics of the problem as a policy issue in the various countries were described.

8.4.1. *Nature of the problem*

The overall image is one of an imminent rather than an acute problem, and also one still beset with scientific uncertainties. However it is recognized by the water supply sectors as a serious problem that deserves adequate attention in order to advert future catastrophes, if not in the environmental then in the financial sense. The paradox is that even successful prevention of new polluting inputs may not solve the problem of historical pollution of aquifers emerging in wells in the coming years. This paradox applies mainly to ground water and less to surface water.

Diffuse source pollution of drinking water resources is not only of agricultural origin. In Great Britain and the Netherlands the water supply sector has dealt successfully with municipalities and the national railways. Industrial pollution stems mainly from point sources, and is often a big problem too.

8.4.2. *Policy issues, policy networks, and policy styles*

Agricultural sector

The three EU Member States offered some fine examples of disintegrating policy communities. All of them have known true agricultural policy communities with iron triangles at their cores for a long time. Developments within the agricultural sector, with the reconstruction of the EU agricultural policy, and the pressure evoked by the general concern about agriculture's impact on the environment as major driving forces, have eroded the bases of the policy communities. The interesting question is whether old disintegrating policy communities are similar to new issue networks concerning the typical policy styles. In the Netherlands and Britain alike, there is a strong external pressure from the European Union to use more direct regulation, now the old ways of the (former) agricultural policy communities have failed to solve the problem of too high nitrate concentrations in groundwater. Perhaps the Netherlands is the most far off from compliance with EU standards, and the Dutch Government is the most concerned. Changes in policy styles are not only towards more regulation (away from ineffective self-regulation), but the consensus base is also diminishing. The big problem is that a new policy must be implemented and enforced within a sector that is accustomed to a consensual approach by tradition, and where support for the new policy is lacking, at least in what the most mobilized parts of the sector are concerned. In the Netherlands in particular, it remains to be seen whether the new top-down approach is feasible, given the resistance and obstruction by certain farmers groups.

Water policy networks

The issue of agricultural water pollution lies in the overlap of two major policy fields: agriculture and water management. The agricultural sector is almost the archetype of a policy community, while the water management sector also has comprehensive policy networks in the countries under study. In Germany and Great Britain, the water supply sector is more or less part of a larger water management sector, which is organized at the national or federal level in all kinds of discussion fora dealing with water pollution, not only from agricultural sources, and the affected interests. Such fora are less manifest in the Netherlands, where the emphasis is on the regional level. Here the water supply sector is relatively well organized around the issue of agricultural pollution of drinking water resources.

8.4.3. *National policy and choice of policy instruments*

There are no specific national policies concerning the issue of protection of drinking water resources. National policies concerning agricultural pollution in general have been developed in the Netherlands and Britain, while the situation in Germany is differentiated, as this concerns the authority of the individual states (see Chapter 7, this volume, in particular Table 7.2). Introduction of new and more stringent regulations of manure practices, evoked by the need of implementation of EU directives, is disrupting the traditional consensual policies, most significantly in the Netherlands. Pesticides policies have been characterized by indirect regulation of pesticide use through regulation of the market and by harmonization all over Europe.

In general the choice of the policy instruments matches the policy network characteristics. To the extent that the traditional policy communities are still functioning, communicative instruments are predominant. The typical policy instruments mix involves information exchange to show farmers sustainable alternatives for the prevailing agricultural practices, and to convince them that these will have no negative effects on their incomes. These communicative instruments are, often supported by economic ones with positive stimuli, such as subsidies for investment or transition costs. The policy instruments applied contain usually no explicit moral appeals to farmers, although the British system of codes of good agricultural practice can be conceived as a way to institutionalize a moderate form of moral appeal.

In all countries the most powerful tool created at the national level seems to be the possibility of establishing some kind of groundwater protection zones. In the Netherlands and Germany this option exists for a relatively long time, the authority is delegated to the Provinces and the States, where it has been applied on a large scale. In both countries it involved mandatory additional restrictions to farmers. In Britain it concerns more recent measures, not widely applied, originally on voluntary and only later on a mandatory base, while the executive power has been kept at the national level, within the Ministry of Agriculture.

Farmers in the groundwater protection zones are entitled to financial compensation to the extent that they are subject to more restrictive regulation than other farmers in the country. In Germany the compensation payment schemes still involve large sums of money. In the Netherlands the differentiation in legal standards was removed out in 1995, as was the right to compensation. In both cases the money was extracted from the water supply companies, and finally from their customers. In Britain only the first cohort of pilot projects was supported by a subsidy scheme. Later on additional legal restrictions were imposed without compensation.

The national policies reviewed are contributing to the protection of drinking water resources in many ways, but the groundwater protection zones are the only direct contributions to the control capacity of the water supply sector.

8.4.4. *Actual control capacity of the water supply sector*

The basic research question concerned the control capacity of the water supply sector. The latter is the designation used for the water supply industry and the regional water authorities. Control capacity is the entirety of:

- the mission and orientation of the organizations of the water supply sector;
- the available organizational resources;
- the selected steering strategies.

Mission and orientation

In all cases, the water supply industry, together with the rest of the water supply sector, endorses the principle of pollution prevention, but they do not consider protection of drinking water resources as part of their core mission. It is a rational choice for them to be actively involved in the protection of drinking water resources, only under certain conditions and up to a certain level. In the Netherlands some water supply companies are remarkably pro-active in developing stimulation programmes, more than elsewhere. This can be explained by the seriousness of the threat of nitrate contamination and, compared to Britain, by differences in the institutional context of the water supply industry.

Available resources and their application

The problem is not that water supply companies are short of financial resources. As far as there are feasible options for investments in reductions of agricultural pollution, they are considering such investments in terms of economic returns. Their involvement has to be of a temporary nature: they invest in bringing about transitions of agricultural practices rather than in structural support of less polluting practices.

The most relevant legal authority for imposing and enforcing additional restrictions on farmers within the water catchment areas is part of the competencies of the water management authorities (NRA, the provinces and the States). Water supply companies do not have legal authority, and they do not want them. In certain cases, however, they have acquired control over land use by buying pieces of land in endangered zones (in Germany and the Netherlands).

The Dutch case study shows the best examples of concerted action by all members of the water supply sector, contributing legal authority as well as financial resources and organizational capacity to preventive activities aimed at farmers. Many of these activities originate from the negotiations about the compensation payments.

Other relevant resources are expertise of farming practices, including sustainable alternatives, acquired confidence and trust in the eyes of farming communities, and information about the state of the water catchment areas. All case studies show several examples of, predominantly small scale, prevention programmes initiated by the water supply sector with reasonable success in terms of farmers' co-operation.

Size is another relevant resource: the scale of the water supply companies does make a difference. Larger companies are able to build up and employ sufficient resources (financial, organizational, legitimacy) to actively deal with the issue. In the Netherlands the larger water companies have taken the lead in dealing with agricultural pollution within their catchment areas.

Strategies adopted by the water supply sector

To the extent that the water supply sector is engaged in preventive activities, they follow in all three cases a consensual approach which involves direct contacts with the local and regional farming communities. Co-operation of the farmers is acquired by persuasion and, to a lesser extent, by material stimuli. (The most important material stimuli are, of course, the compensation payments, but this is no strategic choice of the water supply sector, as the farmers are entitled to it according to the national legislation on groundwater protection zones in Germany and the Netherlands.)

There is a growing distinction between the consensual approach of the water supply sector at the regional and local level and the national trends towards a more regulative policy style concerning agricultural pollution at the national level in all of the three countries. In so far as the national policies are effective, this contributes to the protection of the drinking water resources, but at the same time the legitimacy of environmental policy to the farmers and their confidence in all agencies involved, including the water supply sector, is under pressure. This may have a negative effect on their control capacity.

8.5. Final remarks

The issue of agricultural pollution of drinking water resources as reported here can be characterized as a complex problem with different levels involved, which cannot be considered in isolation.

First, the impact of the European Union is manifest in almost every aspect. Although the main conclusion of the EU case study was that there is very little that the European Union can do directly in dealing with the present problem. However, the European Union has a very important role to play in generating resources and action opportunities for other actors, in the form of legislation and regulations that define the objectives of water quality and set the parameters within which the various groups of actors must operate. The European Union is playing a catalysing role in stimulating actors at the national and regional/local level to integrate considerations of groundwater protection into agro-environmental policies.

Second, at the national level the issue is part of the more comprehensive issue of agricultural pollution in general. Member States are still struggling with the implementation of EU directives concerning water quality standards, while the policy styles towards the agricultural sector are becoming more imposive and regulative, and the traditional agricultural policy networks are under heavy pressure (if they have not been already largely disintegrated). The impact of EU directives on the national level is manifest in making national basic standards of environmental care more stringent. By elevating the reference level, preventive measures of farmers are required without financial compensation. In this way, the EU is having an effect in the member states through command and control policies that imply a compulsory restriction of polluters' choice domain (see Chapter 6, this volume, in particular the taxonomy in Section 6.2.2).

Finally, at the regional and local level, especially where agricultural pollution has damaged drinking water resources most, many successful (but chiefly small-scale) initiatives have been developed by the water supply sector. At the regional and local level, the EU is having an effect on water suppliers that want to go beyond the reference level in water catchment areas by stimulating farmers to more preventive measures. In cases where water suppliers have no regulatory authority, negotiated agreements with farmers, as a specific voluntary approach, appear to be a suitable solution for them to bridge the regulatory gap. In those cases, water suppliers are compensating farmers for overcompliance (see Chapter 6).

The case studies in northern European countries show that rather frequent and direct contacts between the water supply sector and the agricultural sector (inter-relatedness) are an important condition for successfully influencing agricultural practices. This might be a lesson for southern European water policies. The attitude chosen by all parties involved is predominantly rational, oriented at their own (economic) interests. The traditional policy communities at the national level are under heavy pressure, and a joint approach based on common interests and shared perceptions (mutual commitment) is usually not feasible. Indirect steering through the water supply sector (network management) may be an appropriate way to exercise (additional) control over the agricultural sector, especially in those cases where drinking water resources are threatened, but government should keep on performing its control function at some distance and not release the matter altogether.

As already discussed in Chapter 7, negotiated agreements between water suppliers and farmers are an appealing instrument to ease the implementation of agro-environmental targets. However, we wish to stress that the success of such initiatives depends on the combination with command and control policies setting the basic standards of environmental care. The advantage of additional negotiated agreements is not only that they ease the implementation of command and control policies by making preventive action for farmers more financially attractive, they also work as a communicative strategy. The amount of financial compensation paid must be based on information about overcompliance, which means that water suppliers who compensate are buying information about farming practices. This information is useful for monitoring individual compliance and it helps water suppliers to learn more about the agricultural operations in water catchment areas in order to be able to control their water sources.

To draw more lessons from the northern European cases we studied, we will consider them in terms of the common classification of environmental policy instruments as presented in Chapter 6. This classification is underlining the difference between mandatory regulations, economic incentives, and communicative steering (voluntary approaches and negotiated contracts as a specific form).

8.5.1. Regulations

A recurrent theme throughout the whole research was the importance of legal regulation. Legal standards, concerning drinking water quality and also ground and surface waters are virtually the only substantial benchmarks for all parties involved. The research showed also the importance of credible legal regulation: acceptance of the legal standards by policy subjects, and consequent implementation of these standards followed by consequent monitoring and enforcement. These matters are not inconsistent with a policy style aimed at consensus and self-regulation and are, in fact, preconditions to its effective application. In the project, notably the EU legal standards on nitrates and pesticides are spent to be performing this function. Generally speaking the individual member states should not seek by more regulation, but rather through implementation of prevailing EU Directives reinforcement of the control capacity of the water supply sector.

In accordance with the principle of subsidiarity, there is much sense in leaving substantial leeway to the individual member states for specific national and perhaps even regional interpretation of the EU policy and legal regulation. Too rigid restriction imposed upon the member states will harm the policy effectiveness, especially where the governmental relationships with the agricultural sector are problematic and alternatives to the former policy community have to be created.

With respect to new EU regulation, one exception may be made for the establishment of zones with special legal status, such as the groundwater protection zones in the Netherlands and Germany. Examination of the sufficiency of the legal grounds of these very effective policy instruments in EU law and their applicability in all member states may be considered.

Another issue is the distribution of the burden of legal compliance on the parties involved, which is not properly balanced. Farmers are polluting drinking water resources, but it is not easy to make them give account for the consequences of their actions. The prevailing approach of regulating the amounts of manure spread on the land may not be related directly and unequivocally to the resulting environmental damage, but it is one of the few means for getting a legal grip on individual farming practices.

Finally, the credibility and acceptance of EU legal regulation requires ongoing research into their scientific foundations, such as drinking water quality standards, the ecological effects of nitrates and pesticides, especially where these standards are criticized because of their allegedly arbitrary nature.

8.5.2. *Economic incentives*

Positive economic incentives are most effective for influencing farmers behaviour, but the water supply sector is, rightly, opposed to permanent schemes of compensations or rewards for (self-imposed?) restrictions on farming practices. Apart from the issue of the sufficiency of the financial resources, either of the governmental agencies involved or of the water supply companies, these economic incentives may be incompatible with the polluter pays principle. Therefore, positive economic incentives to farmers must not have a permanent nature, and be directed at the stage of transition, such as training, advice, investment costs and, if necessary, temporary income supplements, to take away some barriers for individual farmers who are willing to chance to more sustainable farming methods.

Negative economic incentives are applied in various forms throughout the European Union, notably levies on pesticides use or manure surpluses. The often very modest pesticides levies may serve as transition to the means for registration schemes, but they have no demonstrated effects on pesticides use. The effects of levies on manure surpluses, as applied in the Netherlands, are also unclear, just as the implementation of the measure, including the system of manure bookkeeping, is rather troublesome.

8.5.3. *Communicative steering*

Communicative steering can be undertaken by governments, regional water authorities, or water supply companies. Since several studies have shown that differences in farming practices, notably in the amounts of pesticides used, are not only related to the types of crop or to differences in natural (i.e. geological, hydrological, climatic, etc.) conditions, it makes much sense to address individual farmers on this point.

Communicative steering can be aimed at furnishing knowledge about (more) sustainable agricultural methods, such as effective pesticides use (pesticides leaching into groundwater are also economic losses to farmers), and increasing transparency of individual farming practices, for instance through a system of best management/agricultural practices, to which farmers can commit themselves.

The European Union may consider developing such a system that is attuned at different types of crop, different natural conditions, and perhaps certain national factors. By committing themselves to these standards individual farmers will be able to demonstrate to legal authorities, water supply companies, and to other stakeholders that theirs are sustainable farming practices. Such a system of codification can be complemented by a system of certification (analogous to EMAS). Certificated farmers can recommend their products at auctions and to retailers and consumers by attaching labels to it (which gives them certain competitive advantages over to uncertificated farmers and competitors from outside the EU). Great Britain already has a system of codes of good agricultural practice (without certification) which is very helpful in the relation between farmers and regulators.

With respect to retailers as stakeholders in agricultural practices, notably pesticides use in horticulture, the British case study points to effective interventions by large retail chains (who are serving their own interests by paying attention to consumer attitudes towards pesticides use in horticulture). Similar activities by retail chains are known for other countries.

8.5.4. *Contacts and contracts between the water supply sector and the agricultural sector*

The EU research project (EV5V-CT94-0368) revealed that the situation at the regional and local levels is favourable for effective influence on farmers' behaviour. Good contacts between the water supply sector and farmers promote the latter's receptiveness for communicative steering, while consensual approaches may create favourable conditions for economic and juridical steering.

A common phenomenon at the national level is the disintegration of the old agricultural policy communities. At the same time as the traditional consensual policy style is under pressure, national agricultural policies are changing towards more top-down direct regulation, with all problems of acceptance, control, and enforcement. The project showed that these developments are no obstructions for successful initiatives at the regional and local levels where farmers are addressed through consensual approaches.

The scopes of these initiatives are different. At the national level, it concerns the full agricultural policy, where environmental considerations have been given an important place. The scope of the successful initiatives discussed in the project was much more confined to a particular issue, agricultural pollution of drinking water resources. In terms of policy networks, this concerns second order issue networks, in which some core actors of the larger agricultural policy network participate together with other parties with a stake in the particular issue (and to the extent they have gained access to the policy networks constellations).

Reinforcement of the control capacity of the water supply sector can be realized by the proliferation of these types of second order issue networks. In the first place, this can be aimed for at the horizontal level, to regions with similar problems of agricultural pollution where no direct contacts between the water supply sector and farmers have yet been established.

Proliferation of issue networks may also be considered along vertical lines: reinforcement of the relations between farmers' organizations, water managers and the water supply industry into issue networks on the national and perhaps European level, that should not get entangled in the problems of the agricultural policy networks. Their functions should be to facilitate the proliferation of issue networks on the horizontal level and to exchange information.

References

Bressers, J.Th.A., Kuks, S.M.M., 1992, The agricultural policy network and environmental policy in the Netherlands, Paper presented at the EGPA Conference in Pisa-Italy, September 1992, Enschede: University of Twente, Center for Clean Technology and Environmental Policy.

Bressers, J.Th.A., O'Toole, L.J. Jr., 1995, Networks and water policy: conclusions and implications for research, in: Bressers, J.Th.A., O'Toole, L.J. Jr., Richardson, J.J., eds., Networks for water policy: A comparative perspective, London: Frank Cass, pp. 197–217.

Bressers, J.Th.A., Huitema, D., Kuks, S.M.M., 1995, Policy networks in Dutch water policy, in: Bressers, Th.A., O'Toole, L. Jr., Richardson, J.J., eds., Networks for water policy. A comparative perspective, London: Frank Cass, pp. 24–51.

Dietz, T., Ryecroft, R.W., 1987, The risk professional, New York: Russell Sage Foundation.

European Commission, 1992, The state of the environment in the European Community, EC, Brussels.

Heclo, H., 1978, Issue networks and the executive establishment, in: King, A., ed., The new American political system, American Enterprise Institute: Washington DC, pp. 87–124.

Jehae, M., van Soest, J.P., 1990, Stratego in waterland; bestuurlijke mogelijkheden voor drinkwaterbeleid beperkt (Strategic games in waterland: governance options for drinking water policy are restricted), Nieuwe Bèta, 4, 8–9.

Jordan, G., Schubert, K., 1992, A preliminary ordering of policy network labels, Policy Sciences, 7–27.

Kenis, P., Schneider, V., 1991, Policy networks and policy analysis: Scrutinising a new analytical toolbox, in: Marin, B., Mayntz, R., eds., Policy networks: Empirical evidence and theoretical considerations, Frankfurt/Boulder: Campus Verlag/Westview Press, pp. 25–59.

Kingdon, J.W., 1984, Agendas, alternatives, and public policies, New York: Harper Collins.

Knaap, J.W.M. van der, 1987, Provinciale invloed boven water; ervaringen met de Waterleidingwet in de provincie Utrecht (The provincial influence on water supply; experiences with the Water Supply Act in the province of Utrecht), Bestuur, 10, 300–305.

Kuks, S.M.M., 1988, Mestbeleid in grondwaterbeschermingsgebied: de effectiviteit van bemestingsnormen en schadevergoeding bestuurskundig onderzocht (Manure policy in groundwater protection areas: a policy study on the effectiveness of manure standards and compensation payments). Enschede: University of Twente, Center for Clean Technology and Environmental Policy.

Kuks, S.M.M., Neelen, G.H.J.M., 1991, De vervuiler betaalt: eenzijdig principe in het gebiedsgericht milieubeleid (The polluter pays: a one-sided principle in regional environmental policies). Openbaar Bestuur, 5, 23–28.

Schrama, G.J.I., ed., 1998, Drinking water supply and agricultural pollution: Preventive action by the water supply sector in the European Union and the United States, Dordrecht/Boston/ London: Kluwer Academic Publishers.

Velema, W., Boer, F., Verheul, T., 1989, Aardappelen koken in Spa blauw (Boiling potatoes in Spa blue), Intermediair, 25 (February 24).

VEWIN, 1989, Hoofdrapport en bijlagen Tienjarenplan 1989 (Main report and appendices of the Ten years plan 1989), Rijswijk: National Association of Water Supply Companies.

Waarden, F. van, 1992, Dimensions and types of policy networks, in: Policy Sciences, 29–52.

World Bank, 1992, World Development Report 1992, Development and the Environment, Oxford: Oxford University Press.

CHAPTER 9

Negotiated Agreements on Groundwater Quality Management: a Case Study of a Private Contractual Framework for Sustainable Farming Practices

Marc Barbier and Eduardo Chia

9.1. Introduction

Many economic instruments and regulatory policies have been proposed for addressing negative environmental externalities in order to solve market failures. The issue of externality is not particularly controversial in itself, since it is a well-known theoretical problem.[1] However, the design and the evaluation of economic instruments in real situations are the role of the economist and social scientists in policy making. Among environmental problems, those relating to the effect of farming practices on natural resources are as interesting to analyse as they are difficult to regulate. The issue is basically a matter of combining agricultural and environmental interests in local situations as well as in global policy making.

It is noticeable that European farmers are having to cope increasingly with the interests of other actors in the course of their daily activities. Those actors often want to re-define the objectives and the criteria of good farming practices according to various purposes (economic, social, environmental and political). For farmers and actors alike, the integration of environmental constraints for natural resource protection in farming systems is particularly at stake. This issue is particularly pronounced at the level of water resources management directed towards global European water policy (Barraqué, 1995), since for either quantity or quality management purposes, the effect of agriculture modernization on water resources is a major problem, an one that still has to be resolved.

In the matter of water resources management, as far as France is concerned, during the 1990s, the integration of local environmental constraints and European regulations was particularly slow and contested process. Both the enforcement of the Law on Water of 1992 and the Nitrate Directive have preceded the new Orientation Law for Agriculture of 1999, a law in which the

C. Dosi (ed.), Agricultural Use of Groundwater, 195–212.

definition of a new context for agriculture has clearly taken into account that environmental protection is an objective.

Some French experiences during the 1990s in the field of water management (for example, the Opérations Fertimieux, OGAF Environnement OGAF and Opérations Locales ex-article19) have been analysed in various studies (Barrué-Pastor, 1995; Mormont, 1996; Lemery *et al.*, 1997a), all of which highlight the difficulties of negotiation between farmers' representatives and those gatekeepers who represent the demands put on the environment by society.

Following our claim that economic approaches that focus on the problem of externality could benefit from constructive management science research (Barbier, 1994), in this chapter, we propose to focus on a case study dealing with a process of achieving a negotiated agreement between a private water company (Vittel) and farmers operating in the groundwater catchment area with the aim of improving groundwater quality. The objective of this longitudinal case study (from 1989 to 1997) is to show some practical aspects of the groundwater non-point source pollution problem. According to the key features and implications outlined by Dosi and Zeitouni (Chapter 6, this volume), this case is in some ways a laboratory for negotiated agreement, for several reasons. First, the protection of groundwater, from which the Vittel Company receives mineral water thanks to various springs, has shown the need to treat the catchment area as a continuous space containing heterogeneous sources of nitrogen on-site emissions according to land use by farmers and their agricultural practices in terms of fertilizing. Second, the monitoring of individual nitrogen discharges was not possible without investment in agronomic research at the farm, and it was subject to the willingness and the participation of farmers. Finally, the nitrogen discharges into the groundwater were, in this case, not legally permitted in mineral water (at least, not if the company is selling spring water under the label of natural mineral water).

Bringing about groundwater protection for the Vittel Company's water quality management purposes required a voluntary approach of the type shown by Heinz (Chapter 7, this volume). Nevertheless, such commitments cannot be produced spontaneously by the social actors: the commitment in our case study involved the output of a negotiated agreements which, moreover, implied a process of looking for changes within farming practices and real bargaining between the Vittel Company and farmers involved in nitrogen discharges. As emphasized by Bressers *et al.* (Chapter 7, this volume), in situations of externalities involving a combination of juridical, communicative and economic instruments, other case-studies have been characterized by a high level of interrelatedness and a low level of mutual commitment.

We would like to contribute to the discussion of situations of this kind by focusing on the setting-up of negotiated agreements. Our case study deals with the achievement of such an agreement between the Vittel Company and a number of farmers, an agreement whose purpose was to protect Vittel's catchment area. Not only does this company's investment deserve recognition, thanks are also due to the expertise of a team of researchers from INRA-SAD[2] (Deffontaines and Brossier, 2000). In the light of the success of a specific collective experiment in negotiated agreement, an experiment whose aim was to solve the agricultural non-point sources pollution problem, we would like to present the idea that processes of change in agricultural practice rely on two essential factors: (1) the negotiation of conditions of change by farmers and (2) collective learning in the enactment of transformations that have been negotiated.

The chapter is organized as follows. First, we will show how the process which led to the setting up of a contractual management system unfolded. We will focus particularly on the negotiations and socio-technical changes that occurred during this process. Second, we will describe the collective learning, the conditions under which it took place, and the outcome: an efficient management system for farming practices centred on the improvement of groundwater quality. Third, because researchers were mobilized to enhance negotiation and to 'define better' practices, what they learned from their involvement in such a process deserves to be presented. We will then propose a set of principles for action with a view to facilitating the management of changes which new functions of agriculture and rural space may require. Finally, we try to extract some general lessons from this case study with the view to stimulating a reflexive stance among economists and social scientists in relation to the role they may have to adopt in practical NPS pollution problem solving.

9.2. The negotiation of socio-technical compromises within the management of groundwater quality

9.2.1. *The Vittel case study: an outline*

The Vittel Company produces natural mineral water.[3] In the late 1980s, this company wanted to initiate active protection of the quality of groundwater. It claimed that farmers operating in the catchment area of the spring ought to change their farming systems in order to avoid the discharge of nitrates and pesticides into the groundwater. Similar situations involving harmful discharges are nowadays on the groundwater protection political agenda, but in the late 1980s, it was essentially considered the local problem of a private water company. Because such groundwater resources represent specific assets for water companies like Vittel, heavy investments are an obligatory part of perpetuating the particular quality of these waters and to safeguard their trademark (as the Perrier affair has shown). In respect of this case study, it should be observed that there is no mandatory regulation in France to protect mineral groundwater from potential agricultural pollutant discharges, whereas regulations exist to protect groundwater resources for tap water. The Vittel Company could not find any solution outside a voluntary approach, and was moreover under the

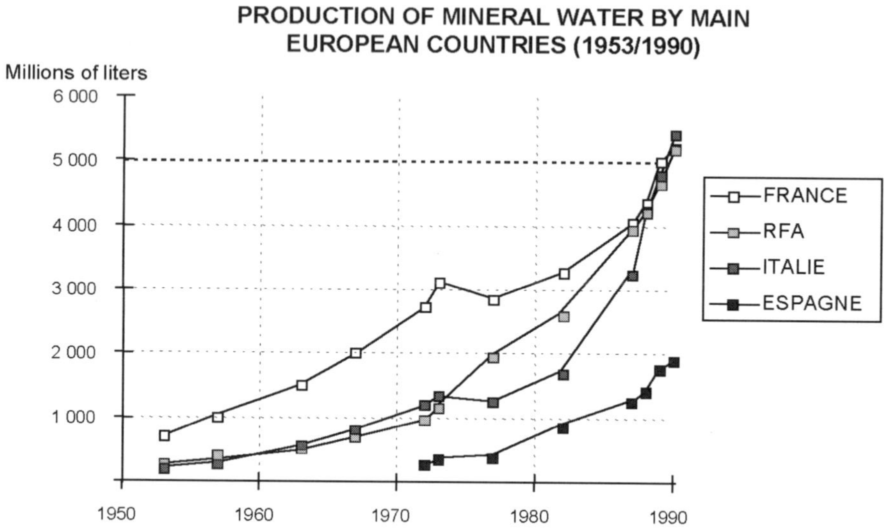

Figure 9.1. The competitive field of natural mineral waters production.

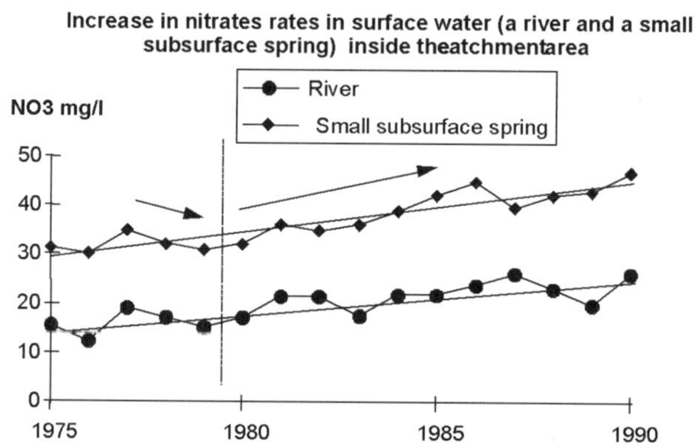

Figure 9.2. Evolution of the quality of surface streams (*source*: Gaury, 1992).

pressure exerted by the high level of international competition in the booming market of mineral waters (Figure 9.1).

As far as Vittel's catchment area is concerned, the purpose of Vittel's managers was then to actively prevent the increase of nitrate concentrations in groundwater since the monitoring of surface streams had already shown such a tendency (Figure 9.2). Active protection thus implied the need to consider a transformation of farming practices within the area delimited by the perimeter of the catchment area, while taking into account its heterogeneity in respect of to soils, topography and crops.

When Vittel managers went public in denouncing the agricultural nitrogen discharges and started to propose buying farmers' lands, an atmosphere of hostility between the farmers' representatives and the Vittel Company was created. Vittel may well have had difficulties in managing such a conflict, since publicizing the risks to groundwater quality represented a major threat to the brand name and possibly would have had legal repercussions.[4] The threat was such that even if Vittel managers had legitimized its claim that change was necessary because of positive impact of its industrial activity in terms of employment and public income tax revenue for the region's economy, it might nevertheless have led to a permanent state of conflict with the farmers. It might also be added that the farmers themselves did not consider themselves responsible for any pollution, since no rules or regulations implied this (Barbier, 1997a). As a consequence, they did not consider the claim for changes promoted by Vittel as obvious, and for many of them, it even seemed unfounded, since in this area, their enterprises were oriented toward the modernization of a production system based on the triptych of milk–mea–cereals.

Although Vittel's aim to change farmers' practices is understandable, their capacity to negotiate far-reaching changes in their production system was not so self-evident. It was not possible for Vittel to impose new practices and new technical systems without the co-operation of the farmers or their representatives. The potential conflict could have had the consequence of greatly damaging the company by making the transformation of the agriculture of the areas impossible, though they were obliged to make the attempt, especially in view of the fact that it would have been completely unrealistic to get rid of the farmers within this area.

Negotiations with farmers thus implied proposing processes for shaping a new system of production geared towards Vittel's objective of limiting the growth of the nitrate rate in the spring water. Vittel thus proposed that a team of researchers from the INRA-SAD should undertake a survey of the existing production systems in order to scientifically assess agricultural pollutant discharges and to propose a technico-economic solution for the sustainable development of local agriculture (Deffontaines *et al.*, 1993; Barbier *et al.*, 1996). Thanks to a multidisciplinary research and development programme, researchers took over the role of scientific experts charged with establishing a diagnosis and advocating the best farming practices to the farmers. This double-hinged role carried with it the possibility that a scientific point of view could bring objectivity into the negotiation framework, a position facilitated by the fact that those INRA researchers treated it under the principles and practices of research action (Chia *et al.*, 1992, 1994). The entrance of scientists into the situation was able to effect a change from a situation of potential conflict to a situation of co-operation while progressively structuring devices for the management of groundwater quality (Raulet-Crozet, 1998).

For a time, the negotiations between the actors concerned focused on goals and means via a search process (Barbier, 1997b). Farmers who had accepted the negotiation framework went from specific problems towards transforming their practices using specific solutions. Their learning progressively took on a certain momentum, a learning that could take place thanks to the setting-up of various framed interactions (Barbier, 1998).

9.2.2. *The negotiation framework between the actors*

A process enabling two or more actors to co-ordinate their actions and finally to come to an agreement supposes at least a minimal common objective (Benghozi, 1990). It implies that actors may exchange their points of view and show what they want, hopefully as inopportunistically as possible. However, in our case study, actors were, in a sense, forced to collaborate because of the legitimate claim by Vittel that agricultural changes were necessary since the potential conflict situation could have tremendous effects on the regional economy. The risk was that opportunism or non-co-operative attitudes could develop defensive routines that would have ruled out compromise.

As Argyris (1995) suggests, organizational routines prevent individuals, groups or inter-groups and organizations from facing the embarrassment of uncertainty and prevent them also from identifying the cause of anti-learning. These defensive routines create obstacles to learning and are over-protective. They depend on the technical and economical constraints on individuals, on cultural aspects and on the representation that individuals may have of the problem. The possibility of going beyond these defensive routines depends on the capacities for negotiating that may develop within individuals. In our case study, the farmers' capacities depended on the financial situation of the farm as well as on the existence of entrepreneurial farming projects. Depending on how much fertilization was required in those projects, some of them might well have been acceptable in view of Vittel's objectives.

The farmers' capacity to negotiate

In the survey of the farmers' attitudes towards Vittel's claim, it appeared that the various negotiating strategies that they developed could be differentiated according to their economic situations and the agricultural project. Given that Vittel demanded a strong constraint, achieving less than 10 mg/l of nitrates under the roots of crops and pasture, for the types of positions that characterized the farmers as shown in Figure 9.3, it appears that for a farm to have a good economic situation did not necessarily entail a good capacity to negotiate with Vittel, since a strong entrepreneurial project could justify opposition to any change. However, farmers in a poor economic situation could be in a position to bargain to adopt a new farming system (Chia *et al.*, 1996).

The socio-economic determinants of farming practices thus intervene in the capacity to negotiate, but they are not sufficient to completely explain the farmers' decisions (Petit, 1981). Indeed, the transformation of production systems that co-operative farmers had to achieve operated in conditions of radical uncertainty for them. In fact, as a result of information on progressively

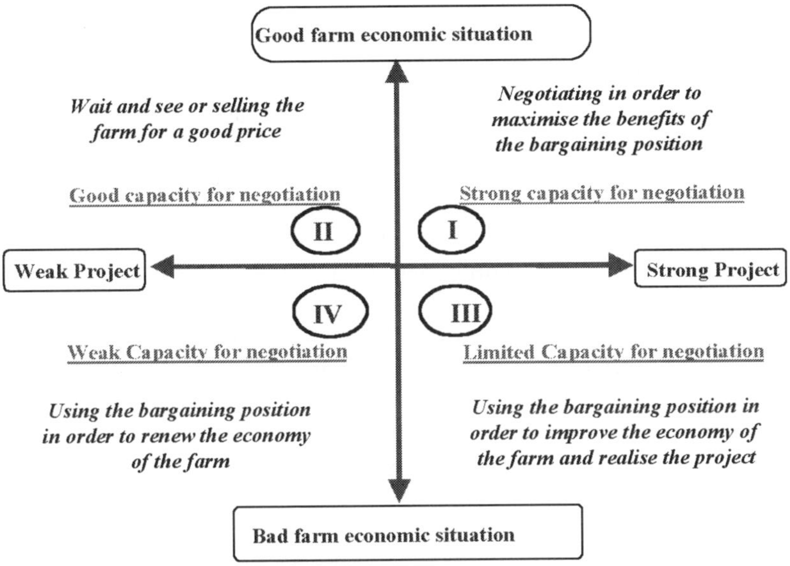

Figure 9.3. Positions of farmers according to their economic situation and their agricultural project.

achieving the technical and economic specifications of a new production system that the researchers gave the Vittel company (Gafsi, 1997), the nature of Vittel's offer changed along with the process. In fact, Vittel's purpose changed towards trying to convince some of the farmers to adopt experimental farming practices in order to find out what the farmers' constraints were and also to show non-co-operative farmers that strong changes were feasible. This is why it appears necessary to us that a second step be taken: considering the dynamics of interactions between various actors in order to understand the progressive shaping of the negotiated agreement as a path-dependent process.

The structuration of management disposals

Farmers' negotiating ability was crucial for asserting their own point of view in the negotiations with Vittel, since the representative from the Agricultural Profession and Farmers' Union kept up a defensive attitude towards the changes that were requested by Vittel. Nevertheless, farmers had to enter a complex two-level set of interactions between Vittel, the Regional Water Agency, INRA researchers and those representatives.

At the first level, a contractual research programme called AGREV formally linked Vittel, the Regional Water Agency and researchers. The aim of this was to assess the local agriculture situation and offer proposals for change. This level involved a co-ordinated process between the researchers and Vittel's managers, as well as frequent contacts with those farmers who had agreed on the principles of change and who were motivated by the idea of carrying out new farming practices to improve groundwater quality.

At the second level, Vittel's managers started to develop what one could interpret as being its own Research and Development project. This project supposed frequent interaction between those managers and representatives of the agricultural Union and of the Agriculture Chamber to implement agronomic and economic prescriptions for new practices, those relationships were supposed to be grounded in the outcomes of the INRA esearch programme. Farmers were actually in a complex set of relationships corresponding to these two levels of ties. They were actors that were participating or reacting to the process of changing the agriculture of the site (Barbier, 1997b).

This double-hinged process of creating a new local agriculture took place within an atmosphere of multiple controversies between actors. One of the important ones was the controversy between the Agriculture Chamber and Vittel in respect of compensatory subsidies for farmers. The time inevitably arrived for Vittel's managers to directly translate scientific propositions into contracts for farming changes with farmers and put aside the ties with their own representatives. At that moment, the institutional representatives of the Agriculture Chamber, the farmers' trade unions and some farmers who had opposed the process, quit the negotiations and became strong opponents (Lemery *et al.*, 1997b).

Each farmer then had to face a process of individual bargaining with Vittel on the basis of a technical specification that had been drawn up by the researchers. Vittel's managers were then in a strong negotiating position with each of them.

The adoption of a negotiated agreement

From 1993 onwards, Vittel's offer to farmers consisted of a set of contracts under the aegis of an 18 or 30 year convention defining liabilities and means of protecting the water quality (Figure 9.4).

In order to sustain structural changes within local agriculture, Vittel created a small enterprise, AGRIVAIR, whose objective was to manage the contractual and technical relationship with the farmers. The function of this enterprise was (1) to internalize those farming practices dealing with manure treatment and application, (2) to reinforce the transformation of farming structures (land, cattle buildings and barns for drying hay) and (3) to aid in fundamental economic re-structuration (supported by dairy quotas). Instead of monitoring and controlling best practices in order to evaluate farmers' commitments to Vittel's objectives, Vittel's managers preferred to internalize the more risky practices of manure treatment and application, ensuring that according to the general convention, chemical fertilizers would not be used.

Given the particular features of this convention, contractual agreements were specific to each farmer and based on their own capacity to negotiate structural changes to farm buildings and land tenure. For their part, the co-operative

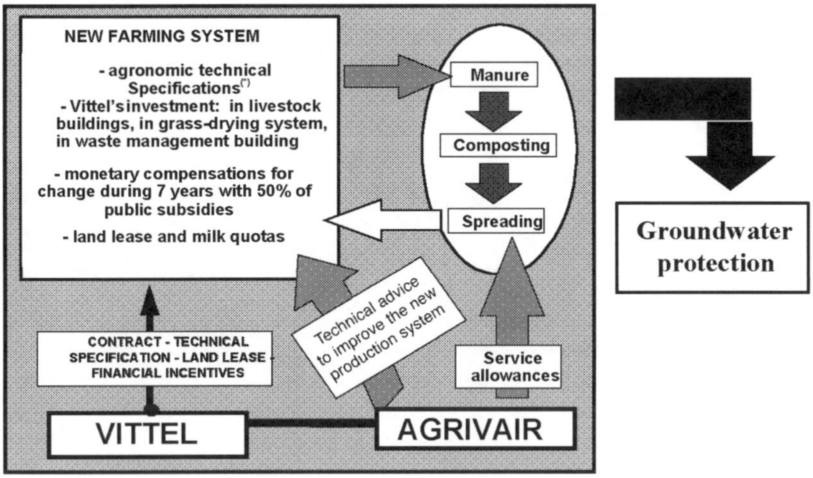

Figure 9.4. The management setting for groundwater protection from nitrogen leaching. (*)Technical specifications: (a) no more maize and pesticides, (b) planned management of manure and spreading, (c) definition of planned fertilizer application, (d) a maximum rate of 1 head of cattle per ha, (e) better farm cropping practices (soil covers and rotation), (f) livestock management with grazing and food diet.

farmers committed themselves to respecting technical specifications (best practices) and thus to providing Vittel with the service of water protection. In order to help each farmer in the process of change, for a transition period of 7 years, some income was provided directly by the Vittel Company and partly by a local organization representing the French Administration and the City of Vittel.

Currently, more than 80% of the farmers in the designated area have such a contract with Vittel. The growth in the number of signatures is shown in Figure 9.5. This growth may be explained in the following way: the contract is financially attractive (the contract foresees a subsidy of around 1500 FF/ha for 7 years), and it includes long-term investments (construction of buildings and manure management). INRA provides valid new techniques (some farmers delayed signing until systems of production and techniques were sufficiently mastered) and, within the farming community, it has an effect at the social level.

The negotiations (even though they have been individual) have been encouraged by a scientific approach and by the presence of researchers on the site for 7 years. These aspects have helped to legitimize Vittel's objectives, particularly for the earlier adopters. The researchers have thus not only shown that the increase of nitrogen in water poses a problem, but have also enabled Vittel to install a management system for water protection. The process should not be seen as a manifestation of a common project, firstly discussed and then collectively implemented.

From the present perspective, it would now seem important to reinforce the

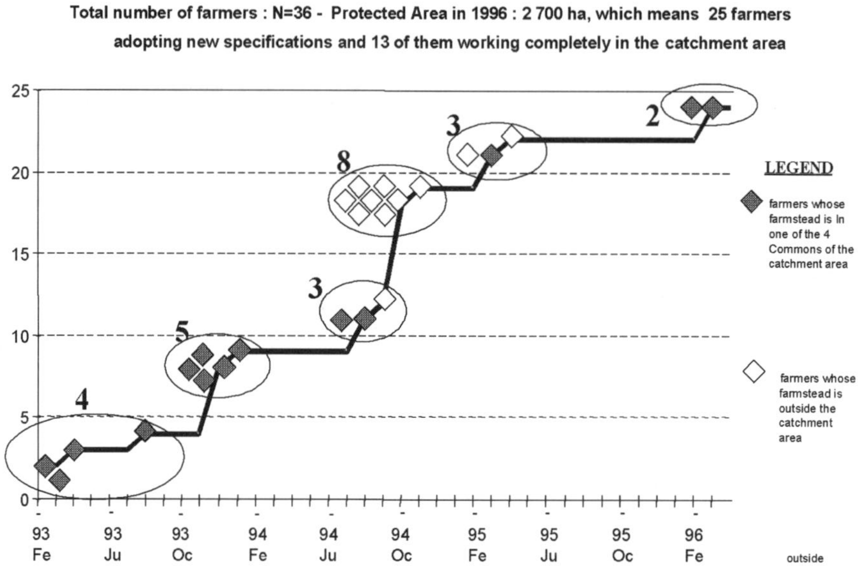

Figure 9.5. The gradual involvement of farmers in adopting the new production system.

collective awareness of this management system in order to guarantee and to strengthen the permanence of local agriculture. This would serve the purpose of improving the organization of the relationship between farmers and AGRIVAIR in order to achieve greater awareness of techniques. Furthermore, it could also enable the contractual setting to be transformed into a resource for local agricultural development, more specifically through marketing of products with green labels from this area.

9.3. From changes to learning

The achievement of a negotiated agreement is the outcome of a process of collective learning undertaken by many actors, from firstly acknowledging the need for radical changes in farming practices to then adapting each farm situation to new best practices. It would therefore seem important to look at how the success of such agreements is reflected in the implementation of contracts and technical specifications on both the sides of the farmers and of Vittel.

9.3.1. *A doubly constrained learning process*

Classically, learning refers to a set of actors inside an organization (firm, administration, workshop, etc.). There are numerous definitions of organizational learning,[5] including that proposed by Koenig (1994): "a collective phenomenon of acquirement and development of skills that, more or less deeply, more lastingly, modifies the management of situations and situations themselves". For those problematic situations where the co-ordination between different actors has to occur within a dynamic context paradoxically characterized by conflictual and co-operative relationships, learning can then be viewed as occurring on two analytical levels.

On one hand, each organization develops knowledge and practices for itself in order to achieve organizational closure (here, especially Vittel, since the farmers' organization did not really see this situation as having the potential for teaching new ways of representing farmers, either for co-operative or non-co-operative purposes). Learning also takes place in between organizations which are actors within a process (in this case, Vittel managers, researchers, farmers and representatives). This kind of learning relies on the development of a set of common ideas about the problem, since no one body is capable of imposing a professional solution for it. For such cases, rather then organizational learning, we would prefer to talk about collective learning (Hatchuel, 1994), since no formal organization or governance structure is there to adjust such a problem.

In this case study, one can thus talk of a doubly constrained learning process, one which, on the one hand, consists of testing the possibility of co-ordinating two professional worlds around a natural resource, and on the other, testing whether productive systems (both those of the water industry and agriculture) can adapt and be transformed. Actors had to modify their common practical knowledge for each stage of the process of change and also had to transform their own work context while taking into account their own experience and the consequence of trials and mistakes. Collective learning also entails a vigilant daily attitude, since actors within the process have to take differences within their own environment into account.

In our case study, the attempt to define a common project marked the first phase of the process. Even though such a project could not ultimately have been undertaken on such a basis, the negotiation terms for the process allowed every actor to provide his own reading of the stakes and in particular, to measure the strength and the determination of the Vittel Company. One can speak of a collective learning which is becoming more and more frequent within new agricultural functions and rural areas in France (Lemery *et al.*, 1996). In the second phase of the process, learning took place directly at the hand of negotiations between farmers, Vittel and its filial AGRIVAIR. The conditions here were change within production systems and within the technical and organizational aspects of these systems at the farm place.

9.3.2. *Learning at the farm place*

The farmers developed some of the skills needed to face up to the complex process of contracting with Vittel. With regard to the setting up of new systems of production, some farmers (in partnership with researchers and with

AGRIVAIR) began to modify their practices (Gafsi and Brossier, 1997). These mainly involved feeding cattle using hay, some techniques for drying this hay and the management of soil fertility on the basis of planned manure spreading. In addition, some farmers developed their own technologies through their ties with local commercial networks, or in the course of creating new networks to meet the technical conditions involved in setting up the new production system.

Nevertheless, the dialogue between the farmers was a sporadic one, and often limited to simple exchanges. It would seem that among farmers, the progression of individual learning towards more collective learning on the basis of the exchange of experiences is quite a tentative one, and there is little sign of a local technical culture starting to develop. Farmers are now observing that some aspects in the domain of agricultural product marketing should have been collectively achieved, notably by developing new merchandizing possibilities and new commercial and technical relationships with downstream firms.

9.3.3. *Learning on Vittel's side*

Learning on Vittel's side was both technical and organizational. At the technical level, the problem for the Vittel's managers was to get suitable knowledge of the local agriculture within the area and a general knowledge of the rules and norms of the agricultural sector. In order to achieve this understanding, Vittel's director hired an agricultural counsellor who had previously worked for the Chamber of Agriculture and who had been a member of the research team for a year. His expertise was then mobilized in the establishment of AGRIVAIR, whose goal was the environmental protection of the catchment area.

AGRIVAIR is in charge of planning the transformation of the production systems of farmers who have signed the contract. It also has a technical role in organizing the production of compost and spreading it, and advising farmers of new techniques. AGRIVAIR's director uses a geographical information system to analyse the changes in the use of land and to localize risky practices. AGRIVAIR is also in charge of co-ordinating the connections between the agricultural profession (Agricultural Administration, Chamber of Agriculture, Technical Institute and Agronomic Research). The importance of AGRIVAIR for the relations between Vittel and INRA should also be highlighted. These have allowed a transfer of expertise from the research program to AGRIVAIR to operate (Barbier, 1998).

9.3.4. *Toward collective learning on the basis of negotiated agricultural agreements*

The absence of a formal group of farmers to negotiate with Vittel and the strategy of individual negotiation that Vittel had promoted led to a logic of assistance and control of agricultural activities within the area (Lemery *et al.*, 1997b). The management system is certainly efficient and functions in accord with Vittel's objective. But a collective approach within local agriculture would need the creation of specific devices to encourage collective learning at the level of the territory. It supposes the development of a technical culture in which farmers would participate while taking in account that farming now has a new function: protecting the environment.

9.4. Some management principles for bringing about the socio-technical transformation of local agriculture

9.4.1. *The role of research in processes of change*

In view of the territorial consequences, the new situation of managing environmental constraints at the farm place imposes on agronomic science the need to go beyond a mono-disciplinary approach that is unfruitful for global solutions. If it is necessary to have interdisciplinary teams of researchers in order to study complex situations, such teams also demand a constructivist attitude toward change, and an attitude supportive of improving direct collaboration between researchers and practitioners (Avenier, 1992). The involvement of researchers in processes of transforming local agriculture and rural spaces thus represents a real challenge for researchers and forces them to abandon linear approaches towards innovation processes (Akrich *et al.*, 1988). In going beyond these approaches, such teams must consider the implications of processes of change according to three dimensions: a comprehensive study of changes, a communicational involvement in the area of negotiation of change, and an advisory position in proposing solutions or frameworks in order to help actors to achieve these changes.

The work of researchers in the field of agricultural changes implies, therefore, a new stance for science: it assumes that the field of expertise has to be widened in order to bring about a reflexive approach towards the consequences (Lemery *et al.*, 1997b; Barbier, 1998). Researchers have then to face multiple and sometimes contradictory missions. On the one hand, they have to develop specifications for new systems of production, to build scientific knowledge about the social, technical and economic conditions of these systems of production, and also to propose analytical frameworks for such situations of agricultural change in order to advocate their use to policy and decision makers. On the other hand, they have to help local actors to develop new perceptions of their activities and to negotiate in order to build new co-ordination frameworks. One way of encouraging the resolution of environmental conflict is surely to increase the actors' learning capacity (Argyris, 1995).

Farmers' advisors, researchers, experts and political mediators[6] (prescripteurs) obviously have to make clear the role they want to play in any process of change they may be involved in. Such an attitude is not a matter of communicational ethics of even a matter of ideology, but a matter of taking a pragmatic approach towards changes that takes account of the conditions and the procedures for bringing about change. For such purposes, discussion procedures have to be designed and co-ordinated (Brossier and Chia, 1994). Research action approaches or participatory research devices can then provide a theoretical basis and lead to the implementation of knowledge in order to achieve further organizational design for the management of processes of change in agriculture.

9.4.2. *Stakes of prescriptions for change*

Although it is difficult to replicate processes of change from a particular situation to another, some methods may be replicated. A way of transferring methods is to identify important stakes and milestones that have to be taken into account by those who intend to prescribe changes, and to facilitate the co-ordination of ways of tackling agri-environmental problems.

The first stake is the instability created by the announcement of the problem, instability increased even more when institutional actors such as researchers or public administration reinforce its legitimacy. This uncertainty has to be mediated through a collective expertise[7] of specific issues that are generated by this uncertainty. It implies specific forms of intervention that are different from a classic scientific expertise aimed at supplying decision-makers with scientific guidelines for action. It means, for example, that researchers have to adopt a maieutic attitude in order to help the various actors involved make sense of the problem.

The second stake relates to the transformations which are necessary to solve the problem. Such transformations correspond to three levels of change: the shaping of new social relationships between actors, the achievement of new economic relations, and the renewal of technical activities. The change is thus as much economic as organizational or technical. For this reason, the design of the intervention may include a management requirement that the controversies be deliberated on and socio-technical compromises found; it is better to publicly express these than keep them hidden.

The third stake relates to the problem of time and space. Intervention for change in agriculture means many actors with diverse positions, many periods of action (or non-action) and many places. Indeed, the degree of actors' involvement will be different according to their objectives and their geographical situations. Actors do not proceed at the same pace. It is necessary, therefore, to avoid slanting the processes too quickly in favour of the fastest actors and those that are more publicly active. One condition is that the transformations can be generalized, and another that early irreversibility be avoided. The points of view of institutions or professional representatives may, for example, betray the broad representation that is normally assumed from them.

The fourth stake relates to the autonomy of the actors' projects. Changes necessarily trigger the problem of instigating new representations and new

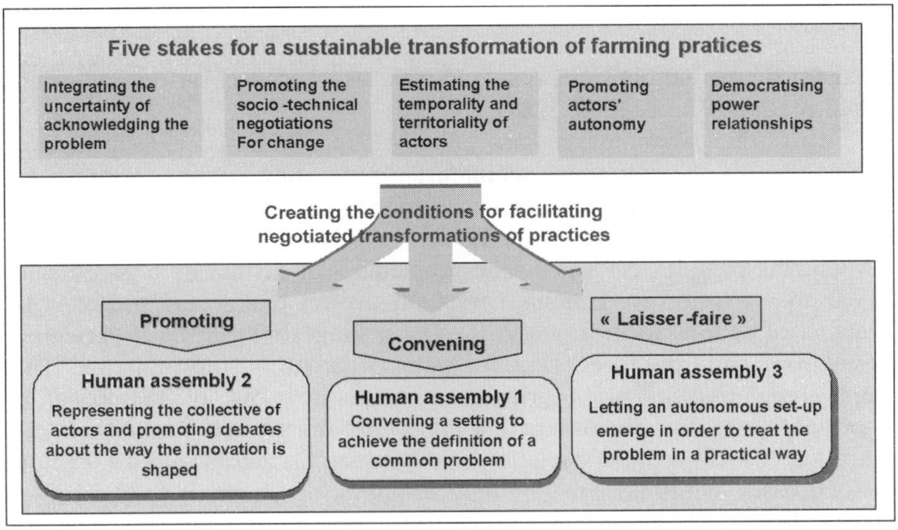

Figure 9.6. A grid to shape and design the management of socio-technical transformations in agriculture.

techniques in daily practices. They should be left to stabilize. It is a question of allowing those who are developing new practices to develop them at their own pace and to allow discussion of specific problems that arise in respect of their practices.

Finally, the fifth stake relates to the devolution of power relationships. The management of the multiple phenomena of power is too often planned according to the model of a technocratic rationale which is supposed to guarantee the effectiveness of changes (i.e. from the point of view of the instigators and according to their own objectives only). Power relationships should not be seen as socially counter-productive aberrations, but as forming part of the situation of negotiation. These relations must therefore be taken into account: this brings with it the possibility of being expressed in the public arena while allowing the expression of actors and the political mediation of their professional representatives.

These five stakes can be dealt with if those who prescribe changes create three types of management set-ups or human assembly (Figure 9.6): (1) a set-up to describe and frame the problem, (2) a set-up to support the emergence of actors who deliberate on and negotiate the organizational design of the innovation process; and finally, (3) a set-up allowing the actors to use the general framework and objective of the process to translate it into new practices.

9.5. Final remarks

The objective of this contribution has been to draw some conclusions from a study of a successful negotiated agreement which included an innovative process and a collective learning situation. Although this case study is obviously unique, it is to some extent very close to a Coasian situation, with bargaining outside a mandatory regulation framework and with transaction costs and asymmetry of information (Coase, 1960, 1988).

Studying the bargaining process underlyining negotiated agreements may help to identify the conditions required fro promoting voluntary approaches since these approaches imply a specific attitude towards the bargaining situation and processes of change (Lévêque, 1997).

Managing complex NPS pollution problems rests neither on a pure regulatory framework nor on a preliminary joint project that would predefine the direction of actions to be engaged in. The manageability of local agriculture for environmental purposes (here, for water protection) is not a matter of the simple regulation of new agricultural techniques, but of the design of co-ordination set-ups, the production of new ordinary practices and the reconfiguration of socio-economic relationships between the actors within a territory. In such cases, contracts are not only economic instruments for promoting change. They are also part of a process of agricultural changes as incentives for learning and as milestones for promoting trust (Chia and Torre, 1999).

Moreover, the process of negotiation which we described in this chapter reveals the importance of the cognitive frameworks of the actors in the framing of the problems to be solved. Actors' capacities to treat them are then not to be considered as only the initial equipment within an economic game but also as a learning resource for the actors in the situation. Negotiation and learning are therefore not as easily separable as a linear conception of decision-making would suggest.

The participation of researchers in such processes of change makes it possible to draw some more general lessons about the ways those who intend to prescribe changes in agriculture should participate. It would seem that the greater the complexity of the situation, the more difficult it is to simplify it and still maintain the separateness of the socio-economic interests of the actors and the biotechnical constraints of their activities right from the start (Coutouzis and Latour, 1986). Indeed, it is precisely at the conclusion of processes of negotiation and learning that such a separation can be produced and realized.

Notes

1. We refer here to seminal paper such as those of Dahlman (1975), Baumol and Oates (1975), Cropper and Oates (1992) and also to Coase (1988), whose approach was crucial for our case study analysis on negotiated agreements.
2. INRA-SAD is a research department of the French National Agronomic Research Institute (INRA).
3. The production of mineral water is regulated by specific orders which distinguish it from other kinds of water because of its "original purety" and of the physicochemical properties that make it beneficial for human health.

4. According to French regulation, mineral water is usually declared originally pure, that is, not under the threat of any pollution risk.
5. This theme of organizational learning is well developed in management sciences and organizational theory; see Argyris and Schön (1978), Fiol and Lyles (1985) and Hubber (1991).
6. The French concept of *prescripteur* has been proposed by Hatchuel (1995) to describe the existence of professions, settings and commercial instruments which prescribe practices, products and service and therefore release the radical uncertainty of agents towards choice in the lack of competencies. The agricultural sector is full of such agents: they play a prescriptive role in relation to farmers' choices.
7. We prefer to use the term expertise rather than experts since we do not believe that such a collective should include scientific experts only.

References

Akrich, M., Callon, M., Latour, B., 1988, A quoi tient le succès des innovations. L'art de l'intéressement, Annales des Mines, Gérer et Comprendre, 11, 4–18 and 12, 14–29.

Argyris, C., 1995, Savoir pour agir. Surmonter les obstacles à l'apprentissage organisationnel, Paris: InterEditions.

Argyris, C., Schön, D.A., 1978, Organizational Learning, Reading, Mass.: Addison-Wesley.

Avenier, M.J., 1992, Recherche-action et épistémologies constructivistes, modélisation systémique et organisations socioéconomiques complexes, Revue Internationale de Systémique, 4, 403–420.

Barbier, M., 1994, Une relecture de la controverse sur le coût social 'COASE vs. PIGOU', où le traitement des problèmes de nuisances ne peut faire l'économie des sciences de gestion, Cahier Lyonnais de la Recherche en Sciences de Gestion, 15, 211–233.

Barbier, M., 1997a, Quand le pollué et le pollueur se découvrent conventionnalistes, Revue Française de Gestion, 112, 100–107.

Barbier, M., 1997b, Water quality protection and technical changes, 51th EAAE Seminar, Zandvoort, Netherlands, April 21–23.

Barbier, M., 1998, Pratiques de recherche et invention d'une situation de gestion d'un risque de nuisance. D'une étude de cas à une Recherche-Intervention, Thèse pour le Doctorat es Sciences de Gestion, Université Jean Moulin – Lyon 3, IAE.

Barbier, M., Benoît, M., Brossier, J., Chia, E., Deffontaines, J.P., Gras, F., Lemery, B., Roux, M., 1996, Programme de recherche Agriculture, Environnement Vittel (AGREV). Rapport de synthèse. INRA, Département Systèmes Agraires et Développement, Unité de Recherche Versailles – Dijon – Mirecourt, 41 pp.

Barraqué, B., 1995, Les politiques de l'eau en Europe, Paris: La Découverte.

Barrué-Pastor, M. *et al.*, 1995, Agriculture, protection de l'environnement et recomposition des systèmes ruraux: les enjeux de l'article 19, Paris: Edition INRA.

Baumol, W.J., Oates, W.E., 1975, The Theory of Environmental Policy, Externalities, Public Outlays and the Quality of Life, Prentice Hall, Englewood Cliffs.

Benghozi, P.J., 1990, La négociation d'une recherche: une étape clé dans la méthodologie d'intervention, Economie et Société, XXIV, Série Sciences de Gestion, 15, 195–209.

Brossier, J., Chia, E., 1994, Participatory research: water quality and changes in farming systems, in: Dent, B., McGregor, M., eds., Rural and Farming Systems Analysis. European Perspectives, London: AB International, pp. 282–304.

Chia, E., Barbier, M., Lemery, B., 1996, Socio-économie du changement, in: INRA-SAD, Rapport final du programme Agriculture-Environnement Vittel, Doc. No. 6, 82 pp.

Chia, E., Brossier, J., Benoît, M., 1992, Recherche-action: qualité de l'eau et changements des pratiques agricoles, Economie Rurale, 208–209, 30–36.

Chia, E., Raulet, N., 1994, Agriculture et qualité de l'eau: négociation et rôle de la recherche. Le cas du programme AGREV, Etudes et Recherches sur les Systèmes Agraires et le, Développement, 28, 177–193.

Chia, E., Torre, A., 1999, Règles et confiance dans un système localisé. Le cas de la production de Comté A.O.C., Sciences de la Société, Presses Universitaires su Mirail, 48, pp. 49–68.

Coase, R.H., 1960, The problem of social cost, Journal of Law and Economics, 3, 1–44.

Coase, R.H., 1988, The Firm, the Market and the Law, Chicago: The University of Chicago Press.

Coutouzis, M., Latour, B., 1986, Le village solaire de Frangocastello, vers une ethnographie des techniques contemporaine, L'année Sociologique, 3, 113–167.

Cropper, M.L., Oates, W.E., 1992, Environmental economics: a survey, Journal of Economic Literature, XXX, 675–740.

Dahlman, C.J., 1975, The problem of externality, Journal of Laws and Economics, 2, 41–163.

Deffontaines, J.P., Benoît, M., Brossier, J., Chia, E., Muller, J.C., Roux, M., 1993, Agriculture et qualité des eaux souterraines. AGREV Diagnostic et propositions sur le site de Vittel. Synthèse et recueil des articles publiés, INJRA-SAD Dijon. 300 pp.

Deffontaines, J. P., Brossier, J., 2000, Système agraire et qualité de l'eau. Efficacité dun concept et construction négociée d'une recherche, Natures, Sciences, Sociétés, 8, 14–25.

Fiol, C.M., Lyles, M.A., 1985, Organizational learning, Academy of Management Review, 10, 803–813.

Gafsi, M., 1997, Ingéniérie d'un processus de changement dans les exploitations agricoles. Cas des modifications de pratiques agricoles pour protéger la qualité d'une eau minérale, Thèse de Doctorat ès Sciences de Gestion, Université de Bourgogne.

Gafsi, M., Brossier, J., 1997, Farm management and protection of natural resources: analysis of adaptation process and the dependence relationships, Agricultural Systems, 55, 71–97.

Gaury, F., 1992, Système de culture et teneurs en nitrates des eaux souterraines, dynamique passée et actuelle en région de polyculture élevage sur le périmètre d'un gîte hydrominéral, Thèse de Doctorat, Ecole Nationale Supérieure Agronomique de Rennes.

Hatchuel, A., 1994, Apprentissage collectifs et activités de conception, Revue Française de Gestion, June/July/August.

Hatchuel, A., 1995, Les marchés à prescripteurs – crise de l'échange et genèse sociale, in: Jacob, A., Verin, H., eds., L'inscription sociale du marché, Paris: L'Harmattan, coll. Logiques sociales.

Hubber, G., 1991, Organizational learning: the contributing process and the literature, Organizational Science, 2, 1–32.

Koenig, G., 1994, L'apprentissage organisationnel: repérage des lieux, Revue Française de Gestion, 97, pp. 76–83.

Lemery, B., Couix, N., Barbier, M., Cartier, S., 1996, Agriculture et environnement: jeux d'acteurs, émergence de nouvelles formes d'organisations et recherche de coordinations sur les espaces ruraux, AIP Nouvelles Fonctions de l'Agriculture et des Espaces Ruraux, Communication au colloque final, Toulouse.

Lemery, B., Soulard, C., Degrange, B., 1997a, Le métier d'agriculteur à l'épreuve des mesures agri-environnementales. Enquêtes dans le département de l'Yonne, INRA-SAD, 72 pp.

Lemery, B., Barbier, M., Chia, E., 1997b, Une recherche-action en pratique: entre production d'eau minérale et agriculture, Etudes et Recherches sur les Systèmes Agraires et le Développement, 30, 71–89.

Lévêque, F., 1997, Voluntary Approaches, Environmental Policy Research Briefs, No. 1, EU Concerted Action on Market Based Instruments

Mormont, M., 1996, Agriculture et environnement: pour un sociologie des dispositifs, Economie Rurale, 236, 28–36.

Petit, M., 1981, Théorie de la décision et comportement adaptatif des agriculteurs, in: Formation des agriculteurs et apprentissage de la décision, Dijon: INRA-INRAP.

Raulet-Crozet, N., 1998, Du conflit à la coopération autour d'un problème d'environnement, Gérer et comprendre, 51, pp. 4–16.

PART IV

Agricultural Policy and Water Use

CHAPTER 10

Agricultural Policy, Environmental Impacts and Water Use Under Production Uncertainty

Giannis Karagiannis and Anastasios Xepapadeas

10.1. Introduction

The effect of agricultural policy on environment is an issue that has been widely addressed both at the theoretical and applied policy level.[1] Price support schemes occupy a central position in the design of agricultural policy, although policies such as production quotas or land set aside schemes are also very important. However, price support schemes, although very popular among farmers, have been associated with increased agricultural pollution. It has been argued that steadily rising farm prices are associated with highly intensive agricultural activities, which in turn progressively require the use of marginal land (agricultural extensification) and the excessive application of various intermediate inputs (agricultural intensification). Both extensification and intensification cause environmental problems. Agricultural extensification brings marginal and highly erodible land into production causing sediment and chemical run-off into surface water. As a result of the conversion of forest, wetlands, and other natural features to agricultural uses, wildlife habitat and the diversity of plants and animals is also affected. Agricultural intensification, on the other hand, tends to increase the use of fertilizers and pesticides, and the concentration of livestock.[2] It may also result in degradation of on-farm resources from salinization, chemical pollution and soil erosion.

It is also well-known that irrigated agriculture puts great pressure on the quantity and quality of water resources in many regions, and that the management of both water quantity and quality in irrigated agriculture has attracted significant attention in agricultural economics (see, for example, Dinar and Zilberman, 1991). Furthermore, where intensive agricultural practices have been adopted, agricultural policies of the price support kind have a profound impact not only on the quantity of water used for irrigation, but also on environmental consequences associated with this use, such as agricultural run-off.

Another major feature of price support is its contribution to eliminating

C. Dosi (ed.), Agricultural Use of Groundwater, 215–239.

output price variability. Production uncertainty remains, however, a significant source of risk in all agricultural activities by being more important in crop rather than in livestock production. More importantly, output variability is closely related to fertilizer and pesticide use, the two most polluting inputs in agriculture. For example, fertilizer increases the probability of very high yields when rainfall is adequate and timely, but also increases the probability of low yields when rainfall is inadequate and chemical burning occurs. Thus, increasing fertilizer use increases both the mean yield and yield variability. On the other hand, pesticide use increases yields when a pest infestation occurs, but does not affect yield if no infestation occurs. By eliminating the lower tail of the yield distribution, pesticide use will increase mean yield and reduce the variance of yield (Leathers and Quiggin, 1991). Seen in this way, fertilizer may be considered to be a risk-increasing input and pesticide a risk-reducing one.[3] However, although this deduction seems reasonable, it is not supported by existing empirical studies. It is evident, however, that in evaluating the environmental impact of agricultural policies, it would be extremely helpful to know whether, for example, fertilizer and pesticides are complements or substitutes or to know what the impact of alternative agricultural policies might be when risk-reducing inputs (water) and risk-increasing inputs (fertilizers) are combined in production processes.[4]

Irrigation, on the other hand, can be regarded as reducing weather uncertainty; water can therefore be considered a risk-reducing input. The combined use of irrigation water with fertilizers generates pollution in the form of agricultural run-off, which in principle can affect surface water bodies or underground aquifers.

The purpose of this chapter is to examine the incidence of two major agricultural policy schemes, price support systems in the form of output price subsidies and land set aside schemes, especially the European land set aside policy, on agricultural pollution and water use. In particular, we will examine the effects of agricultural policy schemes on (i) the use of risk-increasing inputs such as fertilizer and risk-reducing inputs such as water; (ii) agricultural emissions resulting from the combined use of inputs such as fertilizers, pesticides and water; and (iii) the steady-state water head when irrigation water is pumped from groundwater aquifers. We show that when environmental damage from agricultural emissions is not taken into account, then emissions are large relative to the social optimum. This result calls for the introduction of environmental policy and the development of agri-environmental policy schemes. We discuss agri-environmental policy schemes that can correct for environmental externality under given agricultural policy systems.

10.2. Theoretical framework

Following Just and Antle (1990), the major problem in linking agricultural policies, farm practices and environmental problems is that environmental impacts associated with various field level operations do not aggregate to the larger geographical units that are relevant to policy issues. Even though the relationship between agricultural production and environmental damage is site specific (determined mainly by the adopted farm practices), agricultural policies are issued for larger geographical regions. Nevertheless, Just and Antle (1990) have shown that agricultural intensification can be directly linked to environmental damage when the quantity of land in production and land environmental characteristics are held constant. These assumptions are adopted in the following analysis.

It is also assumed that some inputs (fertilizers and pesticides, for example) used by a risk-averse and perfectly competitive farmer may cause environmental problems as well as damage to health. Production is stochastic due to exogenous factors, such as weather conditions (temperature, rainfall), and to the risk–input relationships of productive factors. The level of actual output cannot be known with certainty at the beginning of the production period and farmers have no means to face production uncertainty, i.e. they have no production flexibility to adjust their output ex post. Even though production is uncertain, the cost of production is supposed to be known with certainty and it is measured in terms of planned rather than actual output. The objective of the representative farmer is to maximize the expected utility of profit for any given policy implemented.

Production uncertainty is assumed to have the following form (Just and Pope, 1978): $y = f(x) + h(x)\varepsilon$, where y is the planned output, x is a vector of variable inputs, . is a random variable with $E(\varepsilon) = 0$ and $f_x > 0$, $f_{xx} < 0$ and $h_x > (<)0$ for every risk-increasing (risk-reducing) input. If $h(x) = \text{constant}$, the above model is reduced to a case of additive production uncertainty, which occurs when output variability is independent of the level of planned output. This may be the case for uncertainty arising from weather or other natural phenomena affecting agricultural production. Alternatively, if $f(x) = h(x)$, multiplicative production uncertainty arises, i.e. $y = f(x)(1 + \varepsilon)$. In such cases, the variance of output is positively related to the level of planned output. One example is pest disease where crop loss depends on the level of planned output.

From society's point of view, the existence of an environmental regulator (or a social planner) is assumed. The environmental regulator maximizes the sum of the expected utility of agricultural profits less aggregate environmental damage due to agricultural production. It is assumed that farmers are symmetrical with respect to their risk preference as well as their production behaviour and the regulator takes agricultural policies as given. Environmental damage is captured through a social damage function $D(E)$, which is assumed to be strictly increasing and strictly convex in total emissions E. That is, $D' = \partial D/\partial E > 0$ and $D'' = \partial^2 D/\partial E^2 + 0$, where $E = \sum e_i = ne$, e_i refers to the individual farmer's emission function, and n is the total number of farmers. Given that all farmers are treated symmetrically, there is just one emission function, e, common to all farmers. The emission function is assumed to be a function of inputs used, i.e. $e = g(x_1, x_2)$ where x refers to input quantities. Thus, assuming that x_1 denotes fertilizers and x_2 denotes irrigation water, the emission function denotes the interactions between water and fertilizers in generating agricultural run-off. The emission function is assumed to be strictly increasing and quasi-convex function of input quantities.[5]

The specific method of introducing environmental damage in this paper implies that this damage is imposed not on farmers, but on third parties such as fishermen fishing in a lake with excess phosphorus loadings, or city dwellers using water contaminated by agricultural pollution, or individuals using a polluted lake for recreational purposes. However, agricultural pollution can also affect the farmers themselves if their production function depends not only on the quantity of the water used, but also on its quality. For example, excess salinization of irrigation water caused by agricultural production negatively affects agricultural production that uses the same irrigation water (Xepapadeas, 1996; Dinar and Xepapadeas, 1998). If this type of externality is introduced into the agricultural production function, then strategic interaction might take place among farmers.[6]

10.3. Production subsidies

The objective of a risk-averse farmer is to maximize the expected utility of profit for any predetermined level of production subsidy, s:

$$\max_x Eu(\pi) = Eu\{(1+s)p[f(x) + h(x)\varepsilon] - w'x\} \tag{1}$$

where $u(\cdot)$ is a von Neumann–Morgenstern utility function with $u'(\cdot) > 0$ and $u''(\cdot) < 0$ for risk-averse producers, and p and w are the certain output and input prices, respectively. Problem (1) corresponds to the solution of a private optimization problem, with its solution characterizing market equilibrium for any given exogenous level of production subsidy. In the case of two inputs, with x_1 denoting inputs such as fertilizers or pesticides and x_2 denoting irrigation water, the first-order conditions for (1) are:[7]

$$Eu'(\cdot)\{(1+s)p[f_1(x) + h_1(x)e] - w_1\} = 0 \tag{2a}$$

$$Eu'(\cdot)\{(1+s)p[f_2(x) + h_2(x)\varepsilon] - w_2\} = 0 \tag{2b}$$

On the other hand, the objective of the regulator is to maximize social profit, or the sum of the expected utility of agricultural profit less environmental damage:

$$\max_x Eu(\pi) = n\langle Eu\{(1+s)p[f(x) + h(x)\varepsilon] - w'x\}\rangle - D(ng(x)) \tag{3}$$

with the corresponding first-order conditions for the social optimum:

$$Eu'(\cdot)\{(1+s)p[f_1(x) + h_1(x)\varepsilon] - w_1\} - D'g_1(x) = 0 \tag{4a}$$

$$Eu'(\cdot)\{(1+s)p[f_2(x) + h_2(x)\varepsilon] - w_2\} - D'g_2(x) = 0 \tag{4b}$$

By comparing (2a)–(2b) with (4a)–(4b), we can see that they result in different optimal input use as both $D' > 0$ and $g_i > 0$ for $i = 1, 2$. However, to formally show that price supports without an accompanying environmental policy cause environmental damage which exceeds the socially desirable damage level, it should be proved that the optimal input use at the private optimum, i.e. that resulting from (2a)–(2b), is greater than the optimal input use at the social optimum, i.e. that resulting from (4a)–(4b).

PROPOSITION 1. *At the private optimum, a production subsidy without an accompanying environmental policy will result in an input use that exceeds the corresponding input use at the social optimum. Thus the production subsidy will cause excess environmental damage relative to the socially desirable damage level. This will hold regardless of risk preferences and the risk–input relationship. For proof, see Appendix.*

If we consider two complementary inputs (water and fertilizers, for example), then it is expected that the production subsidy will increase their use relative to the social optimum. This, of course, implies increased emissions relative to the social optimum.

Proposition 1 compares market equilibrium and socially optimal input use when the agricultural policy is fixed. We will now examine the impact from changes in the agricultural policy on input use at the private optimum.

PROPOSITION 2. *Given stochastic separability and complementarity, a similar risk–input relationship and increasing partial relative risk aversion (IPRRA), an output price subsidy will result in an increase in risk-reducing inputs. For proof, see Appendix.*

An intuitive explanation of the above result follows. As output price increases, both expected profit and the variance of profit will increase. As a result, the risk faced by farmers will increase. Under IPRRA, the risk premium increases more than proportionally, and any risk-averse producer seeks to attempt less risk. Since farmers face more risk than would be expected in equilibrium, and given that stochastic complementarity prevails, they will move towards the excessive use of risk-reducing inputs.

10.4. Land set aside

Following Fraser (1991, 1994), the objective function of a risk-averse farmer in the presence of a voluntary set aside programme may be modelled in the following way:

$$\max_{a,x_2} Eu(\pi) = Eu\{p[f(\alpha x_1, x_2) + h(\alpha x_1, x_2)] - w_2 x_2 + r(1-\alpha)x_1\} \quad (6)$$

where x_1: fixed area of land, α: share of land cropped; $1-\alpha$: share of land set aside, r: premium for set aside land per hectare, x_2: water input, and w_2: the unit price of applied water. The first-order conditions for private optimum are:[8]

$$Eu'(\cdot)\{p[f_1(\alpha x_1, x_2)x_1 + h_1(\alpha x_1, x_2)\varepsilon] - rx_1\} = 0 \tag{7a}$$

$$Eu'(\cdot)\{p[f_2(\alpha x_1, x_2) + h_2(\alpha x_1, x_2)\varepsilon] - w_2\} = 0 \tag{7b}$$

For land set aside scheme cases, the optimization problem of the regulator is:

$$\max_x Eu(\pi) = n\langle Eu\{p[f(\alpha x_1, x_2)] - w_2 x_2 + r(1-\alpha)x_1\}\rangle - D(ng(ax_1, x_2)) \tag{8}$$

The first-order conditions for the social optimization problem (8) are:

$$Eu'(\cdot)\{ap[f_1(\alpha x_1, x_2)x_1 + h_1(\alpha x_1, x_2)\varepsilon] + rx_1\} - D'g_1(x)x_1 = 0 \tag{9a}$$

$$Eu'(\cdot)\{p[f_2(\alpha x_1, x_2) + h_2(\alpha x_1, x_2)\varepsilon] - w_2\} - D'g_2(x) = 0 \tag{9b}$$

By comparing (7a)–(7b) with (9a)–(9b), we can see that the conditions for the private optimum result in different optimal input use relative to the social optimum, as both $D' > 0$ and $g_i > 0$ for $i = 1, 2$. However, to show formally that land set aside without a supporting environmental policy causes environmental damage, it has to be proved that the optimal input use resulting from (7a)–(7b) is greater than the input use at the social optimum, i.e. that resulting from (9a)–(9b).

PROPOSITION 3. *At the private optimum, land set aside without an accompanying environmental policy will result in an intermediate input use that exceeds the corresponding input use at the social optimum. Thus the land set aside policy will cause excess environmental damage relative to the socially desirable damage level. This will hold regardless of risk preferences and risk–input relationships. For proof, see Appendix.*

It is not certain, however, that optimal land use under the land set aside scheme is greater than the social optimum.[9] The effects of an increase in set aside premium on the portion of land cropped and on (variable) input use are explored in the next proposition.

PROPOSITION 4. *An increase in set aside premium will increase (reduce) the use of land if land is a risk reducing (risk-increasing) input and reduce (increase) the use of risk increasing (risk-reducing) inputs. This result will hold given stochastic substitutability, different risk–input relationships, and DARA (IPRRA). For proof, see Appendix.*

Thus the effect of changes to the land set aside premium is to increase water use if land is a risk-reducing input. If we assume that a third input (fertilizers, for example) is used in fixed proportion to water, then under the above conditions, an increase in the land set aside premium will increase agricultural emissions.

10.5. Policy implications and the coordination of agricultural and environmental policy

The analysis in the previous sections indicates two basic types of results. The first relates to the comparison of input use between the private and the social optimum for any exogenous given agricultural policy, while the second relates to the effects of changes in agricultural policy parameters on input use.

As has been shown for the two agricultural policies examined, input use at the private optimum exceeds the corresponding input use at the social optimum when no supporting environmental policy is present. Given a strictly increasing relationship between inputs such as water, fertilizers or pesticides and emissions, it follows that without any coordination with environmental policies, the way that agricultural policies are designed will tend to induce agricultural emissions and environmental damage in excess of the corresponding socially desirable levels.

For water use in particular, it should be noted that no opportunity cost for water has been introduced into the social optimization problem. This cost is induced when the depletability of water resources is taken into account by a positive user (scarcity) cost for water. The presence of water scarcity cost, which is not fully reflected in the price of water, will tend to restrict the socially optimal water use relative to the market equilibrium.

The effects of policy changes on emissions are realized through the policy impacts on input use. The total impact on emission is given by the total derivative

$$\frac{de}{d\gamma} = \frac{\partial g}{\partial x_1}\frac{dx_1}{d\gamma} + \frac{\partial g}{\partial x_2}\frac{dx_2}{d\gamma} \tag{11}$$

where $\gamma = \{s, r\}$ indicates the type of applied agricultural policy. Since the theoretical model indicates various signs for the derivatives $dx/d\gamma$, results for the total environmental impact of the agricultural policies under conditions of production uncertainty require that the above theoretical results be coupled with empirical evidence about the risk characteristics of water, fertilizer and pesticide, as well as information about technical inter-relationships (substitutes/complements) among inputs. It is obvious that these relationships may differ across countries, regions and crops due to different weather and soil conditions, water requirements and availability, level of technology, and so on. Thus, conclusive results cannot be drawn solely on the basis of the derived comparative static results.

Unfortunately, there is not much empirical work in this field; more importantly, existing studies of the risk–input relationship for pesticides show different results. Leathers and Quiggin (1991) considered pesticides a risk-reducing input, while Pannell (1991) and Horowitz and Lichtenberg (1994) raised some concerns about the risk–input relationship. Recent empirical evidence shows that fungicides behave as a risk-increasing input in Swiss wheat production (Gotsch and Regev, 1996; Regev *et al.*, 1997). On the other hand, nitrogen and phosphorus have been found to be risk-increasing, while potassium has been found to be risk-reducing (Lambert, 1990; Love and Buccola, 1991; Regev *et al.*, 1997).

Based on the above limited empirical findings, it may be argued that increases in either land set aside premiums or production quotas will result in an increase in fertilizer use. Following Love and Buccola (1991) and Regev *et al.* (1997), this is more likely to occur for nitrogen and phosphorus. On the other hand, if pesticides are considered to be a risk-reducing input, an increase in price support or an increase in land set aside premium will cause pesticide use to increase, while the opposite is true for an increase in the production quota. Similar results hold for potassium, which in some empirical studies has been found to be a risk-reducing input.

Since water is a risk-reducing input, production subsidies will tend to increase water use, while production quotas will tend to reduce it. The effects of land set aside premiums on water use are ambiguous and depend on the land's risk characteristics. If land is a risk-increasing input, then an increase in the land set aside premium will increase water use and vice versa. Thus when the emission function depends on water and say, P or N fertilizers, a change in agricultural policies will have an ambiguous result on emissions since it will change water use and fertilizers in different directions, with the final effects dependent on the form of the emission functions. It should be noted however that even if emissions are reduced, they will still be higher than the socially desirable level if the agricultural policy is not coordinated with environmental policy.

The discussion above suggests that although the sign of the total derivative in eqn. (11) is basically an empirical issue, a comparison between the social optimum and the private optimum indicates that agricultural policies designed in isolation from environmental considerations impose unaccounted for environmental externalities. This suggests the need for coordination between agricultural and environmental policy. This coordination can be brought about in two different ways. In the first, the environmental regulator treats the agricultural policy as fixed and directly introduces environmental policy, which is distinct from the agricultural policy. In the second, it is not possible to directly introduce environmental policy, but the parameters of the agricultural policy can be adjusted to take environmental considerations into account. Thus, the adjustment of the agricultural policy can be regarded as a surrogate for environmental policy.

10.5.1. *Direct environmental policy*

We will assume that the regulator can tax emissions or input use.[10] Let the regulator impose a Pigouvian tax, τ, per unit emissions generated by any farmer. The problem of the private optimum for three policy types examined will be defined in the following way:

(i) Production subsidies

$$\max_x Eu(\pi) = Eu\{(1+s)p[f(x)+h(x)\varepsilon] - w'x - \tau g(x_1, x_2)\} \tag{12}$$

(ii) Land set aside

$$\max_x Eu(\pi) = Eu\{p[f(\alpha x_1, x_2) + h(\alpha x_1, x_2)] - w'x + r_(1-\alpha)x_1 - \tau g(x_1, x_2)\} \tag{13}$$

Comparing the first-order conditions for the above problems with the corresponding conditions for problems (3) and (8), it is evident that

$$\tau = D'(ng(x_1^*, x_2^*))$$

Thus, as expected, the optimal tax equals marginal damage evaluated at the social optimum.

If input taxes are used, then $\tau g(x_1, x_2)$ in eqns. (12) and (13) is replaced by $\tau_i x_i$, $i = 1, 2$, and again by comparing with the first-order conditions for problems (3) and (8) it will be evident that

$$\tau_i = D'(ng(x_1^*, x_2^*)) \frac{\partial g(x_1^*, x_2^*)}{\partial x_i}, \quad i = 1, 2.$$

When we consider irrigation water as an input, this policy calls for a water tax that will reflect the contribution of water to environmental damage.

10.5.2. *Indirect environmental policy*

We will now consider the case in which the regulator cannot introduce direct environmental policy due to, for example, political restrictions, so environmental considerations are introduced by adjusting the agricultural policy parameters.

(i) Production subsidies

The subsidy parameter is perturbed by an amount δs, and the farmer solves the problem

$$\max_x Eu(\pi) = Eu\{(1+s+\delta s)p[f(x)+h(x)\varepsilon] - w'x\} \tag{14}$$

The first-order conditions for problem (14) imply for $i = 1, 2$:

$$Eu'(\cdot)\{(1+s+\delta s)p[f_i(x)+h_i(x)\varepsilon] - w_i\} = 0$$

or

$$Eu'(\cdot)\{(1+s)p[f_i(x)+h_i(x)\varepsilon]-w_i\}+Eu'(\cdot)\{\delta sp[f_i(x)+h_i(x)\varepsilon]\}=0$$

Comparing eqns. (15) and (4), we obtain

$$\delta s=-\frac{D'g_i(x^*)}{Eu'(\cdot)\{p[f_i(x^*)+h_i(x^*)\varepsilon]\}}<0 \tag{16}$$

The adjustment in eqn. (16) is negative, indicating that the subsidy should be reduced in order to take environmental considerations into account. The optimal adjustment equals marginal damage from input use per unit of expected marginal gains from the input use. It should be noted, however, that there is indeterminacy in eqn. (16) unless

$$\frac{D'g_1(x^*)}{Eu'(\cdot)\{p[f_1(x^*)+h_1(x^*)\varepsilon]\}}=\frac{D'g_2(x^*)}{Eu'(\cdot)\{p[f_2(x^*)+h_2(x^*)\varepsilon]\}}$$

This indeterminacy follows from the fact that the adjustment of the subsidy δs can be regarded as an environmental output tax. It is known, however, that the environmental output tax is not optimal when more then one input is used in production (Xepapadeas, 1997). Thus, a second-best subsidy adjustment can be defined in terms of either of the two inputs or some average of the two.

(ii) Land set aside

Let δr be the perturbation of the set aside premium. The farmer will solve

$$\max_x Eu(\pi)=Eu\{p[f(\alpha x_1,x_2)+h(\alpha x_1,x_2)]-w'x+(r+\delta r)(1-\alpha)x_1\} \tag{17}$$

with first-order conditions

$$Eu'(\cdot)\{p[f_1(\alpha x_1,x_2)x_1+h_1(\alpha x_1,x_2)\varepsilon]+rx_1\}+Eu'(\cdot)[\delta rx_1]=0 \tag{18a}$$

$$Eu'(\cdot)\{p[f_2(\alpha x_1,x_2)+h_2(\alpha x_1,x_2)\varepsilon]-w_2\}=0 \tag{18b}$$

Comparing eqn. (18a) with eqn. (9a) we obtain

$$\delta r=-\frac{D'g_1(x^*)}{Eu'(\cdot)x_1}<0 \tag{19}$$

Thus the environmental adjustment will tend to reduce the set aside premium. In such cases, the environmental adjustment acts as an input tax. Since only one of the polluting inputs is taxed, the adjustment is again second best.

The above results indicate that acting as an environmental output tax, indirect environmental policy will tend to reduce production subsidies and, acting as an environmental input tax, also reduce set aside premiums. For the reasons presented above, these adjustments are second-best with respect to the environmental target. The results of Propositions 2 and 4 will determine the effects of the indirect environmental policy on input use. Thus, as water is a risk-reducing input, the indirect environmental policy will tend to reduce its use. On the other hand, however, the same policy will tend to increase the use of risk-increasing inputs. Thus, although the indirect policy is more likely to induce water savings, the final effect of the policy on emissions is ambiguous. This is a result of the second-best character of the indirect environmental policy. This ambiguity calls for closer coordination of agricultural and environmental policy, with environmental policy applied directly through separate instruments.

10.6. The CAP, environmental implications and land set aside

In the EU, the 1992 reform of the Common Market Organization (CMO) introduced a reduction in prices for some major products and two different set aside programmes; rotational and free.[11]

The reduction in prices was expected to reduce the pressure of agricultural activities on environment and eventually reduce emissions. It should be noted that this policy agrees with the theoretical prediction developed in Section 5, which indicated that environmental considerations should be taken into account by reducing the subsidy, as indicated by eqn. (16).

The effects of the set aside programmes can be analysed with the help of the model developed in Section 4. We will take the example of a static case with homogeneous land. According to the EU set aside programme, large producers of cereals, oil seeds and protein crops are given a premium per unit of cultivated land if they set aside a certain proportion of their cultivated land, now set at 10%. If the farmers participate in the programme, they will receive compensation for all their land, both cultivated and set aside. Under the set aside programme, farmers are allowed to grow non-food products on set aside land without losing the set aside premium.

In terms of the model developed in Section 4, under the EU set aside programme and without the farmer growing non-food products on set aside land, the farmer that participates in the programme solves the equation:

$$\max_{x_1, x_2} Eu(\pi) = Eu\{p[f(\alpha x_1, x_2) + (h(\alpha x_1, x_2)\varepsilon] - w_2 x_2 + r x_1\} \quad s.t.\ x_1 \leqslant \bar{x}_1 \tag{20}$$

where x_1: cultivated area of land, $\bar{x}_1$: maximum available land for cultivation, α: share of land cropped which is now a fixed policy parameter, $1-\alpha$: share of land set aside, r: premium for set aside land per hectare, x_2: water input, and w_2: the unit price of applied water.

If it is assumed that the problem has an interior solution[12] for x_1, in the sense of $x_1^* < \bar{x}_1$, then all the conclusions of Propositions 3 and 4 carry over to the EU set aside programme, with a replaced by x_1. Therefore, under the

EU set aside programme, the impact of an increase in the land set aside premium will be an increase in water use if land is a risk-increasing input. There is also, however, the possibility of a boundary solution, in the sense of $x_1^* = \bar{x}_1$. This type of solution will apply under the following circumstances: if the expected marginal product of land is positive up to the maximum available amount of land to the farmers, the farmers will cultivate a proportion a of their whole land, since by doing so they will enjoy an expected positive marginal value product, and at the same time they will receive a premium calculated on the largest possible amount of land. Proof of the boundary solution is shown in the Appendix.

When the total permissible amount of land is used, the amount of water used is the solution to the following:

$$Eu'(\cdot)\{p[f_2(\alpha\bar{x}_1, x_2) + h_2(\alpha\bar{x}_1, x_2)\varepsilon] - w_2\} = 0$$

which is independent of the set aside premium. Since $(\partial x_2/\partial \bar{x}_1) > 0$, a boundary solution for land selection implies increased water use relative to the interior land solution. Thus for a farmer whose land quality is such that he/she expects a positive marginal value product from it, there is an incentive to include all the land in the set aside programme and use the maximum amount of water.

Another aspect of the EU set aside programme is the possibility given to the farmer to use the set aside land for the production of non-food crops, especially those used in the field of renewable resources such as biomass, biofuel, and fibre.[13] Denoting processes associated with the production of food items with f and non-food items with nf, the choice of the inputs for the combined problem of using the specified proportion of land for food production and the set aside part of the land for non-food production can be defined in the following way:

$$\max_{x_1, x_2^f, x_2^{nf}} Eu(\pi) = Eu\{p[f^f(ax_1, x_2^f) + h^f(\alpha x_1, x_2^f)\varepsilon] - w_2 x_2^f + rx_1\}$$
$$+ Eu\{p[f^{nf}((1-a)x_1, x_2^{nf}) + h^{nf}((1-\alpha)x_1, x_2^{nf})\varepsilon] - w_2 x_2^{nf}\}$$

$$s.t. \; x_1 \leqslant \bar{x}_1$$

The optimality conditions for the combined food, non-food productions are:

$$Eu'(\cdot)\{p^f[f_1^f a + h_1^f a\varepsilon + r\} + Eu'(\cdot)\{p^{nf}[f_1^{nf}(1-a) + h_1^{nf}(1-a)\varepsilon\} - \lambda = 0 \tag{21a}$$

$$Eu'(\cdot)\{p^f[f_2^f + h_2^f\varepsilon] - w_2\} = 0 \tag{21b}$$

$$Eu'(\cdot)\{p^{nf}[f_2^{nf} + h_2^{nf}\varepsilon] - w_2\} = 0, \quad x_2^{nf} > 0 \tag{21c}$$

It can be concluded from eqn. (21a) that if the expected marginal value product of the non-food sector is positive, then for a sufficiently small and negative marginal value product of the food sector, the first two terms of the right-hand side of eqn. (21a) will have a positive sum. Then λ is positive and the maximum amount of land is used. So the introduction of the possibility of non-food production in the set aside programme provides incentives for using all available land either for food or non-food production.

The use of land for non-food crops provides an incentive for the production of goods that can be used in the energy sector (cleaner alternatives to traditional fossil fuels, for example). Although this incremental product might generate environmental benefits in terms of fossil fuel saving, eqn. (21c) indicates that, nevertheless it will increase the use of water for the non-food sector. Since the level of water use in the food sector is not affected by the choices in the non-food sector, overall water use will increase as a result of introducing non-food production in the set aside land. According to the emission function, the increase in the overall water use along with the use of the maximum amount of land will increase agricultural run-off. Accordingly, because of the increased water use and agricultural emissions due to cultivation of the set aside land, the decision-making processes will have to take into account that the environmental benefits from fossil fuel savings due to non-food production in the set aside land are less.

6.10.1. *Land set aside and groundwater management*

In the previous sections, water was treated as an ordinary input under the assumptions that farmers were paying a fixed price for water. The effects of agricultural policy on water were viewed in terms of intensity of use as well as contribution to agricultural run-off. In this section, we will explicitly allow the irrigation water to be pumped from a groundwater aquifer and examine the implications of the land set aside policy on groundwater management.

We will assume that irrigation water comes from a 'bathtub' unconfined aquifer with infinite hydraulic conductivity,[14] and is used to irrigate the farms of $j = 1, \ldots, J$ farmers. To simplify the exposition, we will assume that farmers are symmetrical. We will let $H(t)$ be the water table (the height of the aquifer), R a certain recharge rate, β the constant return flow coefficient, and SY the surface area S times the specific yield Y, with the normalization $SY = 1$. The hydrological state of the aquifer will be determined as follows:

$$\dot{H} = \frac{1}{SY}\left[R + (\beta - 1)\sum_j x_2^j\right], \quad SY = 1 \tag{22}$$

It will be assumed that each farmer pumps his/her own water from the aquifer and that the unit pumping costs may be determined in the following way:

$$c(H), c'(H) < 0, \quad c''(H) > 0, \quad \lim_{H \to 0^+} c(H) = +\infty$$

These assumptions introduce decreasing pumping costs in the height of the aquifer and prevent exhaustion. The instantaneous profit for each farmer (assumed in order to simplify the exposition that production uncertainty does not exist), will, under a set aside programme without non-food production, be defined in the following way:

$$\pi^j(t) = pf(ax_1^j, x_2^j) - c(H)x_2^j + rx_1, \quad x_1^j \leqslant \bar{x}_1^j$$

We will assume, as before, that the farmers are small and symmetrical, so that when they choose their water extraction path they will take the extraction paths of the rest of the farmers as given and, furthermore, they will not condition their water extraction on the aquifer height. These farmers are thus following open-loop strategies. The non co-operative open-loop Nash equilibrium for the land use and water extraction under the set aside programme will be defined by the solution to the following problem:

$$\max_{\{x_1^j(t), x_2^j(t)\}} \int_0^\infty e^{-\rho t}\pi^j(t)$$

$$\text{s.t.} \quad \dot{H} = \left[R + (\beta - 1)\left(x_2^j + \sum_{j \neq l} \bar{x}_2^l \right) \right]$$

$$x_1^j \leqslant \bar{x}_1^j$$

The current value Hamiltonian for this problem is defined as:

$$\mathbf{H}^j = \pi^j(t) + \mu^j \left[R + (\beta - 1)\left(x_2^j + \sum_{j \neq l} \bar{x}_2^l \right) \right] + \lambda^j [\bar{x}_1^j - x_1^j]$$

In the Hamiltonian representation of the problem, the costate variable μ^j should be interpreted as the cost of groundwater to the user. Imposing symmetry, the optimality conditions derived from the maximum principle will imply:

$$pf_1(ax_1, x_2)a + r - \lambda = 0$$

$$pf_2(ax_1, x_2) - c(H) + \mu(\beta - 1) = 0$$

$$\lambda(\bar{x}_1 - x_1) = 0, \quad \lambda \geqslant 0 (= 0 \text{ if } \bar{x}_1 > x_1)$$

It is clear that water is pumped up to the point where the marginal value product of water equals unit pumping cost plus the groundwater users cost. The optimality conditions will determine the optimal short-term demand for land and groundwater as being:

$$x_1^0 = x_1^0(a, r, H, \mu, \beta) \quad \text{if } x_1^0 < \bar{x}_1$$

$$x_1^0 = \bar{x} \quad \text{if } \lambda > 0$$

$$x_2^0 = x_2^0(a, r, H, \mu, \beta) \quad \text{if } x_1^0 < \bar{x}$$

$$x_2^0 = x_2^0(\bar{x}_1, H, \mu, \beta) \quad \text{if } x_1^0 = \bar{x}$$

Thus, when there is an interior solution, $x_1^0 < \bar{x}_1$, for land choice, the short-term demands will depend on the set aside parameter a, and water use will depend on the set aside premium r. If we have a boundary solution, $x_1^0 = \bar{x}_1$, and all available land is used,[15] then short-run water demand will solely depend on the height of the aquifer and its user cost. Short-term comparative statics indicate:

1. *Interior solution for land,* $x_1^0 < \bar{x}_1$

$$\frac{\partial x_2^0}{\partial a} = \frac{p^2 a[f_1 f_{12}(1+\varepsilon) - f_{11} f_{22} a x_2^0]}{\Delta}, \qquad \varepsilon = \frac{f_{11} x_1}{f_1} < 0$$

$$\frac{\partial x_2^0}{\partial r} = \frac{-paf_{12}}{\Delta} > 0$$

$$\frac{\partial x_2^0}{\partial H} = \frac{[c' pa^2 f_{22}}{\Delta} > 0$$

$$\frac{\partial x_2^0}{\partial \mu} = \frac{-(\beta - 1)pa^2 f_{11}}{\Delta} < 0$$

$$\Delta = a^2 p^2 (f_{11} f_{22} - f_{12}^2) > 0, \quad f_{12} > 0$$

Thus an increase in the set aside parameter will have an ambiguous effect on water use, while an increase in the set aside premium will increase water use. The effects of the aquifer's height and the user cost on short-term water demand are as expected.

2. *Boundary solution for land,* $x_1^0 = \bar{x}_1$

$$\frac{\partial x_2^0}{\partial a} = \frac{-af_{12}}{f_{22}} > 0, \quad \frac{\partial x_2^0}{\partial H} = \frac{c'}{f_{22}} > 0, \quad \frac{\partial x_2^0}{\partial \mu} = \frac{-(\beta - 1)}{f_{22}} < 0$$

For such cases, an increase in the set aside parameter will increase water use.

These results can be used to characterize the steady-state equilibrium for the height of the aquifer. As shown in the Appendix, under certain conditions, the steady state equilibrium for the aquifer height and its shadow value, defined as (H^*, μ^*): $\dot{H} = \dot{\mu} = 0$, will have a saddle point property. The equilibrium is shown in Figure 10.1. For any initial aquifer height, $H(0)$, there exists an initial shadow value .(0), such that there is convergence to the steady state equilibrium.

The steady-state comparative statics may be obtained by the system:

$$Q \cdot \begin{bmatrix} \dfrac{\partial H^*}{\partial a} \\[2ex] \dfrac{\partial H^*}{\partial a} \end{bmatrix} = \begin{bmatrix} -c'(H^*) \dfrac{\partial x_2^0}{\partial a} \\[2ex] -J(\beta - 1) \dfrac{\partial x_2^0}{\partial a} \end{bmatrix}$$

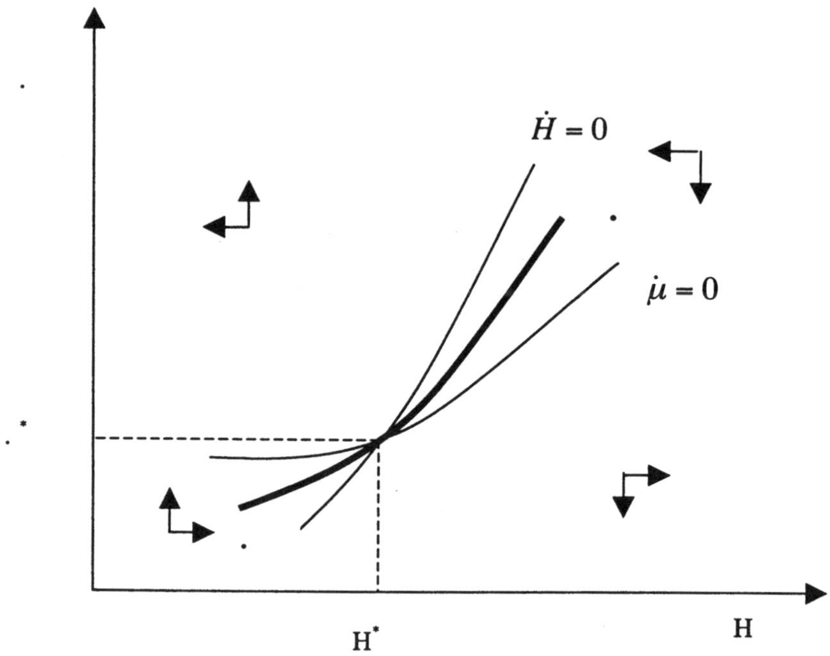

Figure 10.1. Steady state equilibrium for the aquifer height.

$$Q \cdot \begin{bmatrix} \dfrac{\partial H^*}{\partial r} \\ \dfrac{\partial H^*}{\partial r} \end{bmatrix} = \begin{bmatrix} -c'(H^*)\dfrac{\partial x_2^0}{\partial r} \\ -J(\beta-1)\dfrac{\partial x_2^0}{\partial r} \end{bmatrix}$$

The steady-state comparative static results are:

$$\frac{\partial H^*}{\partial a} = -\frac{\dfrac{\partial x_2^0}{\partial a}}{\dfrac{\partial x_2^0}{\partial H}} = \begin{cases} \text{ambiguous} & \text{if } x_1^0 < \bar{x}_1 \\ <0 & \text{if } x_1^0 = \bar{x}_1 \end{cases}$$

$$\frac{\partial H^*}{\partial r} = -\frac{\dfrac{\partial x_2^0}{\partial a}}{\dfrac{\partial x_2^0}{\partial H}} = \begin{cases} <0 & \text{if } x_1^0 < \bar{x}_1 \\ =0 & \text{if } x_1^0 = \bar{x}_1 \end{cases}$$

Thus an increase in the set aside premium r under the set aside programme will not increase the height of the aquifer, while an increase in the set aside proportion a will have ambiguous results if not all the land is used and negative results in the aquifer height when all the land is used.

It should be noted that the qualitative character of these results will not change if we include the possibility of non-food crops on the set aside land.

10.6.2. *A second-best optimum for the aquifer under the set aside programme*

The results obtained above characterize the open-loop Nash equilibrium for the groundwater aquifer. It would be interesting to identify the potential differences when the open-loop Nash equilibrium is compared to a second-best solution.[16] This solution is characterized by the maximization of the sum of a farmer's profits, given the set aside policy. The problem is then:

$$\max_{\{x_1^j(t), x_2^j(t)\}} \int_0^\infty e^{-\rho t}\left(\sum_{j=1}^J \pi^j(t)\right) dt$$

$$\text{s.t.} \quad \dot{H} = \left[R + (\beta - 1)\left(\sum_{j=1}^J x_2^j\right)\right]$$

$$x_1^j \leqslant \bar{x}_1^j$$

The current value Hamiltonian for the second-best problem is:

$$\mathbf{H}^j = \sum_j \pi^j(t) + \mu^{sb}\left[R + (\beta - 1)\left(\sum_j x_2^j\right)\right] + \sum_j \lambda^j[\bar{x}_1^j - x_1^j]$$

If we compare the second-best with the Nash equilibrium, it can be shown[17] that in Figure 10.1, the $\dot{\mu} = 0$ line shifts upwards and the second-best steady-state aquifer height is higher than the open-loop Nash equilibrium height. This discrepancy can be corrected by a water tax which bridges the gap between the open-loop shadow cost of the aquifer height, μ, and the second-best cost, μ^{sb}, with $\mu^{sb} > \mu$. The proposed water tax is fixed and has the property of steering the regulated system to the second-best steady state, although it does not produce the second-best time paths for the state and the costate variables. This steady-state second-best water tax can be defined as:

$$\tau^{sb} = \frac{Jc'(H^{*sb})x_2^{*sb} - c'(H^*)x_2^*}{\rho}$$

Thus managing the groundwater aquifer for the second best situation will require that the set aside policy be accompanied by a water tax.

10.6.3. *The environmental impact*

The environmental impact can be analyzed by taking the example of a situation where withdrawals from the aquifer increase the level of salinity by causing sea water intrusion. This will, in turn, reduce the height of the aquifer. The impact can be introduced into the transition equation for the aquifer site in the following way:

$$\dot{H} = \frac{1}{SY}\left[R + \left(\beta - 1 - s\left(a\sum_j x_1^j, \sum_j x_2^j\right)\right)\sum_j x_2^j\right], \quad SY = 1 \qquad (23)$$

where

$$s\left(a\sum_j x_1^j, \sum_j x_2^j\right)$$

is an increasing and convex function in the aggregated land and water use, indicating reduction in the aquifer's height due to the increase in the level of salinity.

The optimality conditions of the maximum principle for the open-loop Nash equilibrium under symmetry imply:

$$pf_1 a + r - \mu s_1 aJx_2 - \lambda = 0$$

$$pf_2 - c(H) - \mu s_2 Jx_2 - \mu(\beta - 1 - s) = 0$$

$$\lambda(\bar{x}_1 - x_1) = 0, \quad \lambda \geqslant 0 (= 0 \text{ if } \bar{x}_1 > x_1)$$

By comparing these conditions with those derived without any environmental impact, it can be seen that there will be a negative impact on the choice of land through the term $\mu s_1 aJx_2$. This might create an incentive not to use all available land, that is, it might prevent a boundary solution for the land choice. There will also be similar negative impact on the water choice through the terms $\mu s_2 aJx_2$ and $\mu(\beta - 1 - s)$ which will tend to create incentives for water preservation.

It should be noted, however, that if farmers do not consider their impact on salinity; that is, they use the transition eqn. (22) as a constraint for their optimization problem while the true transition is eqn. (23); then there will be excess water withdrawals and salinity will increase relative to the second-best. Along with the set aside policy to correct for the externality, a water tax will be required.

10.3. Final remarks

In this chapter, we examine the impact of such commonly used agricultural policies as production subsidies and land set aside with a special focus on the European set aside policy, on the use of water for irrigated agriculture and on emissions generated by agriculture under production uncertainty.

A comparison of market equilibrium under any given policy with the socially optimal solution under the same policy has shown that agricultural policies without supporting environmental policies will tend to cause input use and environmental damage in excess of what is socially desirable. Thus, excess water use and agricultural run-off will be induced by agricultural policies in isolation from environmental considerations.

An analysis of the effects of changes in agricultural policies on water use has shown that since water is risk-reducing, an increase in production subsidy will tend to increase water use. The result of an increase in the land set aside premium is ambiguous, and will depend on the risk characteristics of the agricultural land. The result will tend to move in the opposite direction for risk-increasing inputs such as nitrogen or phosphorous. Thus the effects of agricultural policy changes on emissions generated by the combined use of risk-reducing inputs (such as irrigation water) and risk-increasing inputs (such as P or N fertilizers) are ambiguous and largely an empirical issue. Due to the lack of empirical findings, the above results should therefore be viewed with caution. These results are in accordance with those related to a riskless world, but they are a lot richer in terms of designing and implementing environmental or agricultural policies for individual inputs.

Our results also suggest the need for coordination between agricultural and environmental policies. If a direct environmental policy in the form of emission taxes or input taxes can be introduced for any exogenous agricultural policy, then the social optimum is attainable.[18] If direct environmental policy is not possible, then agricultural policy parameters can be adjusted. These adjustments are, however, second-best policies and cannot attain the social optimum. In fact, after the adjustment, there is the possibility that water use will be reduced, but emissions increased.

As far as the European set aside policy is concerned, our results indicate that farmers with sufficiently productive land will use the maximum available land and that this policy might therefore increase pressure on water use. It would also seem that use of the set aside land for non-food production increases water use and emissions. This would tend to mitigate the environmental benefits from fossil fuel substitution.

In the dynamic model, increases in the land set aside premium will tend to reduce the aquifer's height in the steady-state equilibrium. A second-best optimum can be achieved by a corrective water tax.

Areas for future research include analysis of the dynamic water model by including a second state variable for the accumulation of pollution such as phosphorus loadings, or the introduction of negative effects from agricultural emissions on agricultural production and strategic interactions among farmers.

Appendix

Proof of Proposition 1: To simplify things, consider that there is only one input used. Given that $D' > 0$ and $g_i > 0$ for $i = 1, 2$, the following inequality may then be obtained from eqns. (2) and (4):

$$Eu'(\cdot)\{(1+s)pF_x - w\}|_{x=x^*} > Eu'(\cdot\{(1+s)pF_x - w\}|_{x=x^0},$$

where $F_x = f_x + h_x\varepsilon$ and x^* and x^0 refer to optimal input choices resulting from eqns. (4) and (2), respectively. The above inequality implies that $x_0 > x^*$ if $F_{xx} < 0$, which is true for a quasi-concave production function (Pope and Kramer, 1979).

Proof of Proposition 2: To analyse the effects from an increase in production subsidy on input use, differentiate eqns. (2a) and (2b) with respect to s and solve the resulting system of equations to obtain:

$$\frac{\partial x_1}{\partial s} = -\left(\frac{p}{|\Delta|}\right)(D_{22} - \delta D_{12})\left\{fEu''(\cdot)\pi_1 + h_1 Eu'(\cdot)\varepsilon + \left(\frac{h}{h_1}\right)Eu''(\cdot)\pi_1 h_1\varepsilon\right\}$$
$$-\left(\frac{p}{|\Delta|}\right)(D_{22}f_1 - D_{12}f_2)Eu'(\cdot) \qquad (5a)$$

$$\frac{\partial x_2}{\partial s} = -\left(\frac{p}{|\Delta|}\right)(\delta D_{11} - D_{12})\left\{fEu''(\cdot)\pi_1 + h_1 Eu'(\cdot)\varepsilon + \left(\frac{h}{h_1}\right)Eu''(\cdot)\pi_1 h_1\varepsilon\right\}$$
$$-\left(\frac{p}{|\Delta|}\right)(D_{11}f_2 - D_{12}f_1)Eu'(\cdot) \qquad (5b)$$

where $\pi_i = p(1+s)[f_i(x) + h_i(x)\varepsilon] - w_i$.

The sum of the first two terms in the brackets of eqns. (5a) and (5b) is positive for risk-reducing inputs and IPRRA (Karagiannis and Gray, 1996; Lemma 2, p. 459).[19] The third term in the brackets is also positive for risk-reducing inputs and decreasing absolute risk aversion (DARA) (Pope and Kramer, 1979; Lemma 3, p. 493). In addition, since $F_{11} < 0$ and $F_{22} < 0$, $D_{22} - \delta D_{12} = pEu'(\cdot)(F_{22} - \delta F_{12}) < 0$ and $\delta D_{11} - D_{12} = pEu'(\cdot)(\delta F_{11} - F_{12}) < 0$ for $F_{12} > 0$ (stochastic complementarity) and $\delta = h_2/h_1 > 0$ (similar risk–input relationships), where $\pi_2 = \delta\pi_1$ and $F_{ij} = f_{ij} + h_{ij}\varepsilon$ for $i, j = 1, 2$. Under stochastic separability $f_1/f_2 = h_1/h_2$, $D_{22}f_1 - D_{12}f_2 = p(f_1 Eu'(\cdot)F_{22} - f_2 Eu'(\cdot)F_{12})$ and $D_{11}f_2 - D_{12}f_1 = p(f_2 Eu'(\cdot)F_{11} - f_1 Eu'(\cdot)F_{12})$, are both negative as long as $F_{12} > 0$.[20] Thus, as $|\Delta| \geqslant 0$, the right-hand side of eqns. (5a) and (5b) is positive.

Proof of Proposition 3: We will assume both that only intermediate inputs cause environmental damage, and that they can be aggregated into a single input. Given that $D' > 0$ and $g_i > 0$ for $i = 1, 2$, the following inequality may then be obtained from eqns. (7b) and (7b):

$$Eu'(\cdot)\{pF_x - w\}|_{x=x^*} > Eu'(\cdot)\{pF_x - w\}|_{x=x^0},$$

where x^* and x^0 refers to optimal intermediate input choices resulting from eqns. (9b) and (9b), respectively. The above inequality implies that $x^0 > x^*$ if $F_{xx} < 0$, which is true for a quasi-concave production function (Pope and Kramer, 1979).

Proof of Proposition 4: Differentiate eqns. (7a) and (7b) with respect to r and solve the resulting system of equations to the following:

$$\frac{\partial \alpha}{\partial r} = \left(\frac{x_1}{|\Delta|}\right)\{D_{22}[Eu'(\cdot) - (1-\alpha)Eu''(\cdot)\pi_1] + D_{12}(1-\alpha)Eu''(\cdot)\pi_2\} \quad (10a)$$

$$\frac{\partial x_2}{\partial r} = \left(\frac{x_1}{|\Delta|}\right)\{D_{12}[Eu'(\cdot) - (1-\alpha)Eu''(\cdot)\pi_1] + D_{11}(1-\alpha)Eu''(\cdot)\pi_2\} \quad (10b)$$

Given Lemma 2 by Pope and Kramer (1979, p. 493), the bracketed term in both eqns. (10a) and (10b) is negative (positive) as $h_i < (>) 0$, and $D_{12} < 0$ if $F_{12} > 0$ (Hiebert, 1982). On the other hand, given Lemma 2 by Karagiannis and Gray (1996, p. 459), the bracketed term in eqn. (10a) and eqn. (10b) is positive (negative) as $h_i < (>) 0$, and $D_{12} < 0$ if $F_{12} > 0$ (Hiebert, 1982).

Boundary solution for the EU land set aside programme
If $[f_1(ax_1, x_2) + h_1(\alpha x_1, x_2)\varepsilon] > 0$ for $x_1 \in [0, \bar{x}_1]$ then $x_1^* = \bar{x}_1$.
The first-order conditions imply:

$$Eu'(\cdot)\{p[f_1(\alpha x_1, x_2)a + h_1(\alpha x_1, x_2)a\varepsilon] + r\} - \lambda = 0$$

Under the assumption of uniformly positive marginal product of land, this equality is satisfied only if $\lambda > 0$. But the Kuhn-Tucker conditions would then imply that $x_1^* = \bar{x}_1$

Proof of the Saddle Point Equilibrium
From the maximum principle, the Modified Hamiltonian Dynamic System (MHDS) characterizing the dynamic behavior of the aquifer height and the use cost will be defined in the following way:

$$\dot{\mu} = \rho\mu + c'(H)x_2^0$$

$$\dot{H} = R + J(\beta - 1)x_2^0$$

Let (H^*, μ^*): $\dot{H} = \dot{\mu} = 0$ denote the steady state of the groundwater system. We can then characterize this steady state under the assumption that unit pumping costs are nearly linear at the steady state or $c''(H) \approx 0$. The slope of the isocine for μ will be defined in the following way:

$$\left.\frac{d\mu}{dH}\right|_{\dot{\mu}=0} = -\frac{c'\dfrac{\partial x_2^0}{\partial H}}{\rho + c'\dfrac{\partial x_2^0}{\partial \mu}} > 0$$

while the slope of the isocine for H will be defined as:

$$\left.\frac{d\mu}{dH}\right|_{\dot{H}=0} = -\frac{\dfrac{\partial x_2^0}{\partial H}}{\dfrac{\partial x_2^0}{\partial \mu}} > 0, \quad \left.\frac{d\mu}{dH}\right|_{\dot{H}=0} > \left.\frac{d\mu}{dH}\right|_{\dot{m}=0}$$

The isocine for H then cuts the isocine for μ from below and a unique steady state is determined as shown in Figure 10.1.

The Jacobian determinant of the MHDS will be defined in the following way:

$$|Q| = \begin{vmatrix} \rho + c'\dfrac{\partial x_2^0}{\partial \mu} & c'\dfrac{\partial x_2^0}{\partial H} \\ J(\beta - 1)\dfrac{\partial x_2^0}{\partial \mu} & J(\beta - 1)\dfrac{\partial x_2^0}{\partial H} \end{vmatrix} = \rho J(\beta - 1)\frac{\partial x_2^0}{\partial H} < 0$$

and the steady state will have the saddle point property shown in Figure 10.1.

Notes

1. See, for example, Bonnieux and Rainelli (1988), deWit (1988), Tobey and Reinert (1991), Abler and Shortle (1992), LaFrance (1992), Plantinga (1996), Lewandrowski *et al.* (1997), CAP (1977).
2. By using a panel data set encompassing 23 countries (including the EU) over the period 1982–87, Lewandrowski *et al.* (1997) provided empirical evidence for a positive and significant effect of price support on fertilizer use per hectare, but they found weak statistical evidence for the effect of price support on land use.
3. According to Pope and Kramer (1979), an input is said to be marginally risk-reducing (risk-increasing) if, under risk aversion, the expected value of its marginal product is less (greater) than marginal factor cost. Consequently, a risk-averse farmer uses less (more) of a marginally risk-increasing (risk-reducing) input under production uncertainty than otherwise.
4. Leathers and Quiggin (1991) and Karagiannis (1998) have recently examined the interaction between agricultural policy and the environment under conditions of production uncertainty and risk. In particular, they considered the impact of output and input price subsidies on fertilizer and pesticide use. Their analyses are, however, subject to certain limitations: first, they are restricted to a single-factor framework which does not consider the technical inter-relationships among inputs, and second, they focus solely on price policies, neglecting the environmental impact of other commonly used schemes, such as land set aside.
5. Thus, $g_i = \partial g/\partial x_i > 0$ and $g_{ii} = \partial^2 g/\partial x_i^2 > 0$ and $g_{ij} = \partial^2 g/\partial x_i\,\partial x_j > 0$ for $i \neq j$. The assumption on the cross partials implies that an increase in the level of irrigation water used will increase marginal emissions, for example P-loadings, from fertilizers.

6. This approach could be a fruitful area of further research since it allows for a more realistic repesentation of the interactions between the natural water system and agricultural production. For a methodological approach to this problem and the implication of strategic interactions, see Xepapadeas (1997).
7. The second-order conditions are satisfied under risk aversion and a quasi-concave production function.
8. It is assumed that the second-order conditions are satisfied.
9. This can be seen from the inequality, $Eu'(\cdot)\{pF_x - w - rx\}|_{x=x^*} > Eu'(\cdot)\{pF_x - w - rx\}|_{x=x^0}$ implied by eqns. (9a) and (7a), where x^* and x^0 refer to optimal land use at the social and the private optimum respectively. Even with $F_{xx} < 0$ it is not certain that $x^0 > x^*$ as $rEu'(\pi) > 0$.
10. We disregard here the well-known problems associated with the non point-source pollution character of agricultural pollution (see, for example, Xepapadeas, 1997, Chapter 4).
11. See CAP (1997).
12. The optimality conditions for this problem are similar to those of eqn. (6), with a replaced by x_1, and the additional condition, that at the solution x_1^*

$$\lambda(\bar{x}_1 - x_1^*) = 0, \quad \lambda \geqslant 0 \; (= 0 \text{ if } \bar{x}_1 > x_1^*)$$

where . is the Lagrangean multiplier associated with the constraint $x_1 \leqslant \bar{x}_1$. Similar conditions also hold for the regulator's optimization problem.
13. We do not analyse the implications of long-term environmental set aside in order to create biotopes or small natural parks.
14. For definitions, see Koundouri (1999a,b).
15. This requires $f_1 + r > 0, \forall (x_1, x_2)$.
16. We call this solution second best because we consider the agricultural set aside policies as fixed and not derived through a particular optimization procedure.
17. The result is obtained by comparing the MHDS for the second-best problem

$$\dot{\mu}^{sb} = \rho\mu^{sb_2} + Jc'(H)x_2^0$$

$$\dot{H} = R + J(\beta - 1)x_2^0$$

with the corrresponding MHDS for the open-loop problem defined in the Appendix.
18. In principle, with perfect competition in the permits market, similar results can be obtained using tradeable emission permits.
19. A sufficient condition for IPRRA is that relative risk aversion is non-decreasing and absolute risk aversion is non-decreasing in profit, with at least one of them being non-constant (Katz, 1983). IPRRA indicates that a proportional increase in risk results in a more than proportional increase in aversion to risk.
20. Stochastic separability, defined as $\partial(F_i/F_j)/\partial\varepsilon = 0$, for the Just–Pope production function implies that $h_i(f_j + h_j\varepsilon) - h_j(f_i + h_i\varepsilon) = 0$ or $f_i/f_j = h_i/h_j$ for $i \neq j$ (Karagiannis, 1997). This holds as long as all factors of production have similar risk–input relationships. Thus, by definition, stochastic separability holds under multiplicative or additive production uncertainty.

References

Abler, D.G., Shortle J.S., 1992, Environmental and farm commodity policy linkages in the US and the EC, European Review of Agricultural Economics, 19, 197–217.

Bonnieux, F., Rainelli, P., 1988, Agricultural policy and environment in developed countries, European Review of Agricultural Economics, 15, 263–280.

CAP Working Notes, 1977, Agriculture and Environment. Special Issue, European Commission, Luxemburg.

de Wit, C.T., 1988, Environmental impact of the CAP, European Review of Agricultural Economics, 15, 283–296.
Dinar, A., Xepapadeas, A., 1998, Regulating water quantity and quality in irrigated agriculture, Journal of Environmental Management, 54, 273–289.
Dinar, A., Zilberman, D., eds., 1991, The Economics and Management of Water and Drainage in Agriculture, Boston: Kluwer Academic Publishers.
Fraser, R., 1991, Nice work if you can get it: An analysis of optimal set-aside, Oxford Agrarian Studies, 19, 61–69.
Fraser, R., 1994, The impact of price support on set-aside responses to an increase in price uncertainty, European Review of Agricultural Economics, 21, 131–136.
Gotsch, N., Regev, U., 1996, Fungicide use under risk in Swiss wheat production, Agricultural Economics, 14, 1–9.
Hiebert, L.D., 1982, Production uncertainty and factor demands for the competitive firm: an extension, Southern Economic Journal, 48, 211–215.
Horowitz, J.K., Lichtenberg, E., 1994, Risk-reducing and risk-increasing effects of pesticides, Journal of Agricultural Economics, 45, 82–89.
Just, R.E., Antle, J.M., 1990, Interactions between agricultural and environmental policies: A conceptual framework, American Economic Review, 80, 197–202.
Just, R.E., Pope, R.D., 1978, Stochastic specification of production functions and economic implications, Journal of Econometrics, 7, 67–86.
Karagiannis, G., 1997, Production uncertainty, risk aversion and cost minimization, Greek Economic Review, 19.
Karagiannis, G., 1998, Interaction between agricultural and resource policy: note, American Journal of Agricultural Economics, 80, 419–421.
Karagiannis, G., Gray, R., 1996, The LeChatelier principle under production uncertainty, Scandinavian Journal of Economics, 98, 453–460.
Katz, E.., 1983, Relative risk aversion in comparative statics, American Economic Review, 73, 452–53.
Koundouri, P., 1999a, A survey of the economics of groundwater resources: Gisser–Sanchez's results reconsidered, mimeo, University of Cambridge.
Koundouri, P., 1999b, A dynamic model of sectoral adaptation to resource depletion with backstop technology: theory and application to groundwater allocation, mimeo, University of Cambridge.
LaFrance, J., 1992, Do increased commodity prices lead to more or less soil degradation?. Australian Journal of Agricultural Economics, 36, 57–82.
Lambert, D.K., 1990, Risk considerations in the reduction of nitrogen fertilizer use in agricultural production, Western Journal of Agricultural Economics, 15, 234–244.
Leathers, W.D., Quiggin, J.C., 1991, Interactions between agricultural and resource policy: the importance of attitudes toward risk, American Journal of Agricultural Economics, 73, 757–764.
Lewandrowski, J., Tobey, J., Cook, Z., 1997, The interface between agricultural assistance and the environment: chemical fertiliser consumption and area expansion, Land Economics, 73, 404–27.
Love, H.A., Buccola, S.T., 1991, Joint risk preference-technology estimation with a primal system. American Journal of Agricultural Economics, 73, 765–774.
Magnusson, G., 1969, Production under risk: a theoretical study, Uppsala: Almqvist and Wiksells Boktryckeri AB.
Pannell, D.J., 1991, Pests and pesticides, risk and risk aversion, Agricultural Economics, 5, 361–383.
Plantinga, A.J., 1996, The effect of agricultural policies on land use and environmental quality, American Journal of Agricultural Economics, 78, 1082–1091.
Pope, R.D., Kramer, R.A., 1979, Production uncertainty and factor demands for the competitive firm, Southern Economic Journal, 45, 489–501.
Regev, U., Gotsch, N., Rieder, P., 1997, Are fungicides, nitrogen and plant growth regulators risk-reducing? Empirical evidence from Swiss wheat production, Journal of Agricultural Economics, 48, 167–178.

Tobey, J., Reinert, K., 1991, The effects of domestic agricultural policy reform on environmental quality, Journal of Agricultural Economics, 43, 20–28.
Xepapadeas, A., 1996, Quantity and quality management of groundwater: an application to irrigated agriculture in Iraklion, Crete, Environmental Modelling and Assessment, 1, 25–35.
Xepapadeas, A., 1997, Advanced principles in environmental policy, Cheltenham UK: Edward Elgar Publishers.

CHAPTER 11

Agricultural Subsidies, Water Pricing and Farmers' Responses: Implications for Water Policy and CAP Reform

Javier Calatrava and Alberto Garrido

11.1. Introduction

Agricultural policies have major impacts on other sectoral policies such as water or environmental policies. In Mediterranean countries with large irrigation acreages, agricultural policy is one of the most important factors affecting irrigation water demand. The kinds of subsidies, barriers to trade or market interventions in place largely dictate what irrigated crops farmers are willing to grow and to what extent farmers are likely to respond to various water pricing policies. While most OECD countries, with the European Union at the forefront, are progressing towards charging water consumers for the costs they impose on the water supply systems, many of them still promote irrigation projects, seek market protection for certain crops, and establish generous grandfathering provisions for irrigators' obligation to pay higher water rates (OECD, 1999; Garrido, 1998; Dinar and Subramanian, 1997).

The side-effects of pursuing policies with mutually offsetting effects have not been fully recognized. For instance, the objective of making irrigators responsible for at least the operation and management costs of the water supply can be effectively challenged on the grounds that the government still bears the larger burden of the costs of new irrigation projects whose beneficiaries will not be made responsible for the entire supply costs. There is a similar inconsistency in irrigators getting subsidies to invest in new irrigation equipment, while the government sets the objective of promoting and protecting certain agricultural commodities that are mostly grown in irrigated areas. It may be argued that if the government wants to charge farmers higher water rates, it is perhaps reasonable to help them improve their ability to pay for water, lowering other farming costs such as capital costs for investments in better irrigation technologies, or insuring higher commodity market prices. However, two sources of inefficiencies may result from such carrot and stick measures. One is the cost of implementing a two-sided policy requiring enforcement and monitoring of

C. Dosi (ed.), Agricultural Use of Groundwater, 241–257.

two separable areas, namely, agricultural markets and water use. The other stems from the fact that unless farmers are given fixed water allowances, they may end up consuming more water if the demand effect is larger than the deterrent effect via higher water rates. Furthermore, if there is sufficient encouragement to adopt better technology, or if water demand functions have a certain elasticity in their levels, farmers may respond accordingly and perhaps end up using more water than with the previous technology (Huffaker and Whittlesey, 1995; Garrido *et al.* 1997, Huffaker *et al.*, 1998).

In this paper, we investigate the effects of various agricultural policy scenarios on farmers' demand for water. Specifically, the paper will analyse the quantitative differences in the effects of two possible policy scenarios; namely, the current agricultural subsidies of the CAP and the one implied by Agenda 2000. It also attempts to determine how various water pricing policies could affect the allocation of agricultural land to crops currently subject to maximum guaranteed quantities or reference surface, and resulting levels of nitrogen percolation. Lastly, it will determine the extent to which agricultural policies can mitigate the negative income effects on farmers of water pricing policies in line with the Water Policy Framework Directive, as it has currently been drafted.

We will apply a slightly modified version of a mathematical programming model developed by Garrido (1998, 2000) and Calatrava and Garrido (1999) to simulate farmers' behaviour in the presence of several agricultural programmes, and use the simulated results to carry out a set of regression analyses. After reviewing the literature on farmers' water demand and how agricultural policies affect it, the chapter proceeds with a theoretical discussion about the effects on protection mechanisms for various farm products. The fourth section is devoted to presenting the empirical model and the context in which it will be applied, namely, irrigated agriculture in the south of Spain. There will be some concluding remarks.

11.2. Literature review

Cummings and Nercessiantz (1992) were among the first authors to cast doubts on the possibility of using water-charging policies to improve water use efficiency. Apart from the legal impediments, their negative judgement is based on the assumption that water right holders may be doubly penalized for increases in water charges, as they may have bought the land for a price in which cheap access to water may have already been capitalized. Gardner (1997) supported these conclusions. Brill *et al.* (1997) showed that passive trading, a policy alternative by which the management sets a price, and respects the three-legged criteria of Fundenberg and Tirole (efficient allocation, balanced budget, equity-rent distribution in proportion to historical water use), is better than active trading under costly information requirements and transaction costs, and block-rate pricing.

Other studies on water demand for irrigation seem to suggest that its elasticity is highly sensitive to the level of water charges and a number of institutional and policy factors. Iglesias *et al.* (1998) have shown that situations which have relatively similar pricing policies exhibit wide disparities in the effects they create, pointing out that major initial conditions such as water allowances and the state of the conveyance facilities largely determine the quantitative differences across irrigation districts. A remarkable inelasticity in irrigation water demand functions at low charges has been found by writers who include Schaible (1997), Varela-Ortega *et al.* (1998), Montginoul and Rieu (1996a,b) and Moore *et al.* (1994). Beare *et al.* (1998) analysed various rating options in which seasonal components operate as consumption deterrents during peak-demand periods. Their results show that the value of water is highly sensitive to the factor of access to transportation facilities, suggesting that resource pricing needs to be accompanied by access pricing by facilities.

Rainelli and Vermersch (1998) showed the effects of having a common agricultural policy for French irrigated agriculture and pointed out the big influence that a national programme aimed at supporting farmers' investments in irrigation equipment had on the growth of farmers' water consumption in France. These authors conclude that part of the recent water scarcity problems experienced in France was due to the increase in irrigated agriculture, that in turn was promoted by the above mentioned policy.

11.3. Framework for the model

11.3.1. *A model of an individual irrigator (production payment system)*

We will let an irrigator's optimization problem be given by:

$$\max \pi(W_k, S_k) = \Sigma_k S_k[P_k + EP_k]F^k(W_k) - \Sigma_k C_k(S_k) - p_w \Sigma_k W_k S_k$$
$$s.t. \quad \Sigma_k S_k W_k \leqslant W_0 \quad \text{and} \quad \Sigma_k S_k \leqslant S_0 \tag{1}$$

where
$\pi(.)$ = farm's profit function, $F^k(W_k)$ = production function of crop k with partial derivatives $F^k_w > 0$, $F^k_{ww} < 0$, $C_k(S_k)$ = cost function of crop k, with $C' > 0$ and $C'' > 0$, W_k = per hectare amount of water applied to crop k, S_k = surface devoted to crop k, P_k = price of output k, EP_k = CAP payment to crop k, p_w = institutional charge on water consumption, W_0 = per hectare water allotment for the farm, S_0 = farm's surface.

With the above mentioned curvature properties, eqn. (1) yields an optimum solution ($W^*_1, \ldots, W^*_K, S^*_1, \ldots, S^*_K, \lambda^*, \mu^*$), where W^*_k, and S^*_k, denote the optimal amount of water applied to crop k and the surface devoted to crop k, respectively; and λ^* is the shadow price of water, while μ^* is the shadow price of land.

The Lagrangian function for eqn. (1) is given by:

$$L = \Sigma_k S_k\ [P_k + EP_k]F^k(W_k) - \Sigma_k C_k(S_k) - p_w \Sigma_k W_k S_k - \lambda[\Sigma_k S_k W_k - W_0] - \mu[\Sigma_k S_k - S_0] \quad (2)$$

Taking derivatives with respect to W_k and S_k, the following necessary conditions are obtained:

$$\lambda^* = [P_k + EP_k]F^k_w - p_w \quad \forall k \quad (3)$$

$$\mu^* = [P_k + EP_k]F^k - C'_k - p_w W^*_k - \lambda^* W^*_k \quad \forall k$$
$$\text{or } \mu^* = [P_k + EP_k]\ F^k - C'_k - W^*_k [P_k + EP_k]F^k_w \quad \forall k \quad (4)$$

The dual variable of the optimal solution associated with water (λ^*) allows us to elicit a water demand function. If we parameterize W_0, we will then obtain different solutions to eqn. (1), including different values of shadow prices of water. We will let $\Psi(.)$ be the function that relates water and its shadow price or model 1's dual value, expressed as follows:

$$\lambda^*(W) = \Psi(W; p_k, p_w, EP_k, S_0) \quad (5)$$

which can be thought of as an inverse water demand schedule, since it links water availability with the irrigator's willingness to pay for water. Using comparative statics analysis based on eqn. (5) it can be shown that $\partial\lambda^*(W)/\partial W \leqslant 0$, indicating that the water demand is negatively sloped (Garrido, 1995).

11.3.2. *A model of an individual irrigator (area payment system)*

The irrigator's optimization problem would now be given by:

$$\max \pi(W_k, S_k) = \Sigma_k S_k [P_k F^k(W_k) + ES_k] - \Sigma_k C_k(S_k) - p_w \Sigma_k W_k S_k$$
$$\text{s.t.} \quad \Sigma_k S_k W_k \leqslant W_0 \quad \text{and} \quad \Sigma_k S_k \leqslant S_0 \quad (6)$$

with the same notation as in the previous case.

The Lagrangian function for eqn. (6) is:

$$L = \Sigma_k S_k [P_k F^k(W_k) + ES_k] - \Sigma_k C_k(S_k) - p_w \Sigma_k W_k S_k - \lambda[\Sigma_k S_k W_k - W_0] - \mu[\Sigma_k S_k - S_0] \quad (7)$$

Taking derivatives with respect to W_k and S_k, the following necessary conditions are obtained:

$$\lambda^* = P_k F^k_w - p_w \quad \forall k \quad (8)$$

$$\mu^* = P_k F^k + ES_k - C'_k - p_w W^*_k - \lambda^* W^*_k \quad \forall k,$$
$$\text{or } \mu^* = P_k F^k + ES_k - C'_k - W^*_k P_k F^k_w \quad \forall k \quad (9)$$

If we look at expressions (3) and (8), we notice that in the second one, the shadow price for water does not depend on the value of the subsidy payment to each crop, but only on the value of the marginal product of water for each crop and the institutional charge on water. For the production payment scheme, the shadow price for water also depends on the value of the CAP payment to each crop.

This suggests that agricultural subsidies have a direct influence on the water demand of farmers for the case of a production payment system, such as the one existing in the EU up until the 1992 CAP Reform. The rate of change of the shadow price of water in response to changes in the amount of price support of a given crop *k* will be given by the value of the marginal product of water for that crop.

In the case of the current area payment system, a change in the subsidies seems to have no influence on the shadow price of irrigation water, or, for that matter, on the demand for water, but that does not mean that it has no influence on the level of water consumed. In any of these two cases, the price of water (we have assumed a proportional pricing system) has an influence on $\partial\lambda^*/\partial ES_k$ or $\partial\lambda^*/\partial EP_k$, that is, on the effect of a change in the subsidies on water demand. In both cases, the marginal effect of a change in the price for water over its shadow price is constant and equal to minus one ($\partial\lambda^*/\partial P_w = -1$).

From this theoretical point of view, there seems to be no interrelation between the individual effects of agricultural subsidies and water pricing over farmers' demand for irrigation water. It seems logical that as long as subsidies affect farmers' revenue, pricing policies will influence their costs.

11.4. Methodology

An empirical application of the above analysis was carried out using a non-linear mathematical programming model simulating farmers' decisions on land and water allocation in two water districts located in the Guadalquivir River Basin in southern Spain. This empirical version of model (1) was calibrated to each representative type of farm in each water district and adapted to allow for an analysis of the consequences of CAP subsidies and water pricing on farmers' demand for water in those areas.

Various exogenous variables in the model were parameterized in order to set the different scenarios to be considered in the analysis. These parameters are:

- CAP subsidies scenarios: two different sets of crop prices and area payments were considered, one corresponding to those of the 1997–1998 season, and the other to those considered in the Agenda 2000 Berlin agreement.
- Cotton price: two different prices for cotton were established, one corresponding to the price paid in years of excess production, and subsequently low prices, and the price for normal years (a higher one).

- Set aside: four different levels of set aside were considered (0%, 5%, 10% and 15%).
- Water charges: apart from the usual maintenance, management and energy costs paid in each irrigation district, an extra institutional price for water was included in the model. Following the idea of an equal price for all districts, regardless of their location and water availability, this institutional price was set at 0, 3, 6 and 10 Spanish pesetas/m^3 (0, 0.018, 0.036 and 0.06 euros/m^3).

Combining the values considered for each of the above parameters, 64 different orthogonal scenarios were obtained. Each district's model was run for these 64 scenarios and for a series of 86 different water allotments, ranging from zero to 8500 m^3/ha. This was to obtain pairs of points of water availability and the corresponding dual value, making a total of 5504 runs of the model for each water district.

Apart from the dual value for water, for each scenario and level of water availability, other results were obtained. These include the corresponding level of nitrogen percolation per hectare and the percentage of farm area devoted to each crop.

We then performed a regression of each of these results on water availability levels and on the corresponding value for each of the parameters. The idea was to determine the quantitative and qualitative effects of these various parameters on the shadow price of water, the nitrogen percolation, and the surface devoted to various crops that are supported by one or another set of policies. The equations pertain to each of the four districts considered in the study, and the data used for each district's regression results from an aggregate of the results of each individual representative farm.

When the results are analysed, it must be taken into account that the input data for the regressions does not come from observed reality. The dependent variables are output from the mathematical programming optimization models, while the explanatory variables are the values of the parameters considered in each model run. As a consequence, explanatory variables are not correlated at all, which allows their individual effect on the explained variable to be isolated. On the other hand, this also means that any interpretation of the regression results should be taken as merely indicative, even though the math models have been calibrated to the real conditions of each district. The variables used for the regression analysis are explained in Table 11.1.

11.5. Results and discussion

We will look first at the inverse water demand equation, in which the shadow price of water is the dependent variable. Results for both districts are reported in Table 11.2. All variables except SETASIDE are highly significant and the

Table 11.1. Variables definition in the regression analysis

Variables in regression analysis	Notation in the theoretical model	Meaning	Values and units of measurement	
SHADWATER	λ	Shadow price of water (or, dual value of water)	$\geqslant 0$	(Ptas/m^3)
LWAT	W_0	Logarithm of the water availability	>0	(Log(cubic meter/pha))
WATCHARGE	P_w	Institutional water charge	0	ptas/m^3
			3	ptas/m^3
			10	ptas/m^3
				ptas/m^3
AG2000	EP_k or ES_k	Agricultural policy scenarios	1	For Agenda 2000
	P_k		0	For current CAP reform
SETASIDE	(not in the theoretical model)	Set aside requirements for cereals, oil and protein crops	0	% of set aside land
			5	% of set aside land
			10	% of set aside land
			15	% of set aside land
COTPRICE	P_{cot} (not explicitly present in the theoretical model)	Subsidised price of cotton	1	Current price/ Guide EU price
			0	Low support price/ Minimum EU price

Table 11.2. Inverse water demand equations (dependent variable: the shadow price of water, SHADWATER)

Coefficients	Irrigation district Bembezar (ID-B)	Irrigation district Fuente Palmera (ID-FP)	Irrigation district Guadalmellato (ID-GU)	Irrigation district Genil Cabra (ID-GC)
INTERCEPT	88.29 (111.54)	73.74 (58.24)	97.48 (84.24)	81.32 (73.32)
LWAT	−8.48 (−89.29)	−8.31 (−54.69)	−10.71 (−77.12)	−9.29 (−68.82)
AG2000	1.66 (9.55)	5.53 (19.9)	2.9476 (11.59)	5.82 (23.56)
SETASIDE	−0.0035 (−0.23)*	−0.0102 (−0.41)*	−0.00875 (−0.38)*	−0.0031(−0.14)*
COTPRICE	40.33 (231.9)	40.29 (144.9)	29.59 (116.43)	39.77 (161)
n. of obs.	5504	5504	5504	5504
Adjusted R^2	0.9785	0.9571	0.9352	0.9582

(t-ratio in brackets). (*) not significant.

model captures a high percentage of the variation in the shadow price of water. The level (in logs) of water availability LWAT is equally significant and provides a clue about the water demand elasticity. The AG2000 coefficient is positive and highly significant, and implies that the Agenda 2000 provisions for the EU common agricultural policy may shift the water demand functions outward. This would have two major implications for water policy. One is that if water savings are sought from the changes envisioned in the Water Framework Directive, water charges should increase more with Agenda 2000 than with the agricultural policies put in place with the 1992 CAP reform. The other is that farmers would be willing to pay more for the water with Agenda 2000 than is the case with the 1992 CAP reform; hence, water charging policies can be more effective in raising more revenue from agricultural water users, which was identified as one major objective of the Water Framework Directive.

However, in the Guadalquivir Basin, the effect of the Agenda 2000 provisions may be offset by future changes in the Cotton Common Market Organization. Currently, the cotton industry in the European Union is supported by a minimum price and guide price which is well over the international adjusted price of non-ginned cotton (Garrido and Mesquida, 1998). Although there is no international trade for non-ginned cotton, European farmers receive a price for their harvests which effectively doubles the international price. The regression results show that if the cotton support system is removed, the water demand functions would shift inwards. Thus the Agenda 2000 effects may be completely offset by reductions in the level of support delivered by CAP to the European cotton producers. Finally, the set-aside provisions do not seem to have a profound impact on the shadow price of water.

Figures 11.1 and 11.2 depict the water demand functions for all districts, each figure corresponding to one scenario (*Scenario 1*: low water charge (3 ptas/m^3) + CAP reform as defined in 1992; *Scenario 2*: low water charge (3 ptas/m^3) + Agenda 2000). Common assumptions for all curves are a set aside

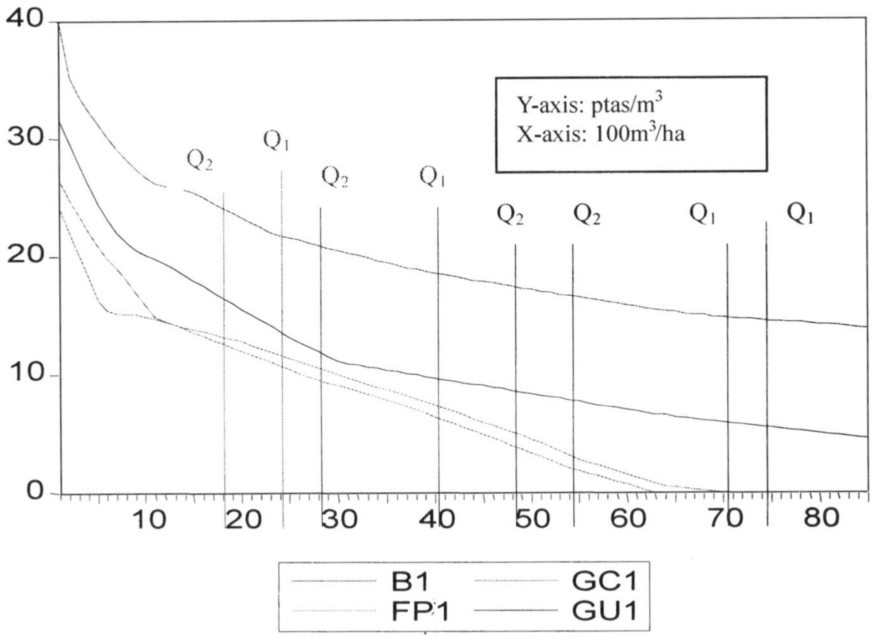

Figure 11.1. Inverse water demand functions for the four districts (pre Agenda 2000 payments).

of 5% of the land devoted to COP crops, and a low level of cotton price support. The vertical lines drawn over the shadow price curves represent the current normal allotment (Q_1) and a reduction of about 30% (Q_2). Figures 11.3 and 11.4 compare water demand functions for scenarios 1 and 2 for Bembezar and Fuente, Palmera and Genil Cabra, and Guadalmellato, respectively.

The curves show that water demand functions are quite elastic. These results go against the literature on irrigation water demand functions. This may be due to two complementary factors: one is that crop cost functions are convex, whereas most of the other studies use linear cost functions; and secondly, yield functions are non-linear and concave, which allows for a wide variety of production techniques for crops which can easily grow under rain-fed conditions as well as optimal water applications, such as wheat, sunflowers and sugar beets (grown in the winter season in Andalusia). Both the yield and cost functions curvatures provide much more flexibility than Leontieff technologies. Note, however, that our model is static and does not allow for changes in irrigation technology. Hence, a long-run demand function would perhaps exhibit an even greater elasticity.

The curves depicted in Figures 11.1 and 11.2 exhibit some differences across various farming situations. The most noteworthy are due to the effects of the Agenda 2000 scenario. In the Bembezar and Guadalmellato districts, the shift

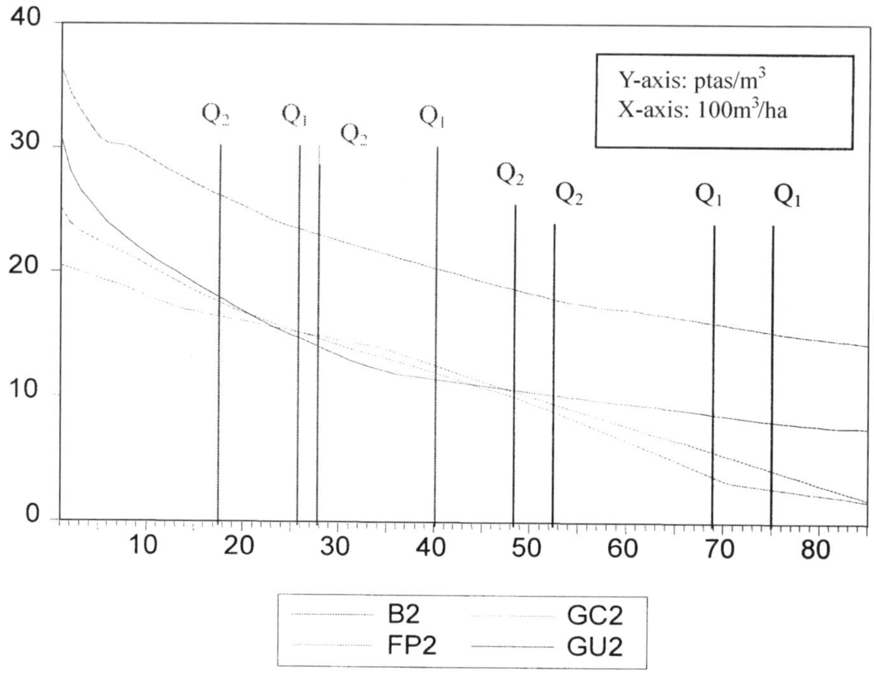

Figure 11.2. Inverse water demand functions for the four districts (Agenda 2000 scenario).

in the inverse demand functions due to the implementation of Agenda 2000 is smaller than in Fuente Palmera and Genil Cabra. This has major implications for water demand management and water pricing policies. In addition to this, the option to exchange water rights that the Spanish legislative bodies passed in December 1999 has added an increasing level of policy uncertainty on top of the profound changes that Agenda 2000 and the Water Framework Directive will impose on the current water users.

By looking at the various shadow prices of water within various policy scenarios and water allotment situations, we can anticipate the market situations shown in Table 11.3.

Table 11.4 reports the regression results of the equation that has acreage devoted to cotton as the dependent variable. Both the availability of water (WAT, here not in logarithmic terms) and the Agenda 2000 scenario (AG2000) are positively related to the cotton acreage. The latter result is an indication that farmers would probably shift from COP crops to cotton as a result of the changes that Agenda 2000 will impose on these crops, although this result is conditional on the level of support provided to the cotton sector after implementation of Agenda 2000. Interestingly, however, the price of water does not seem to influence the cotton acreage very much, indicating that farmers would keep

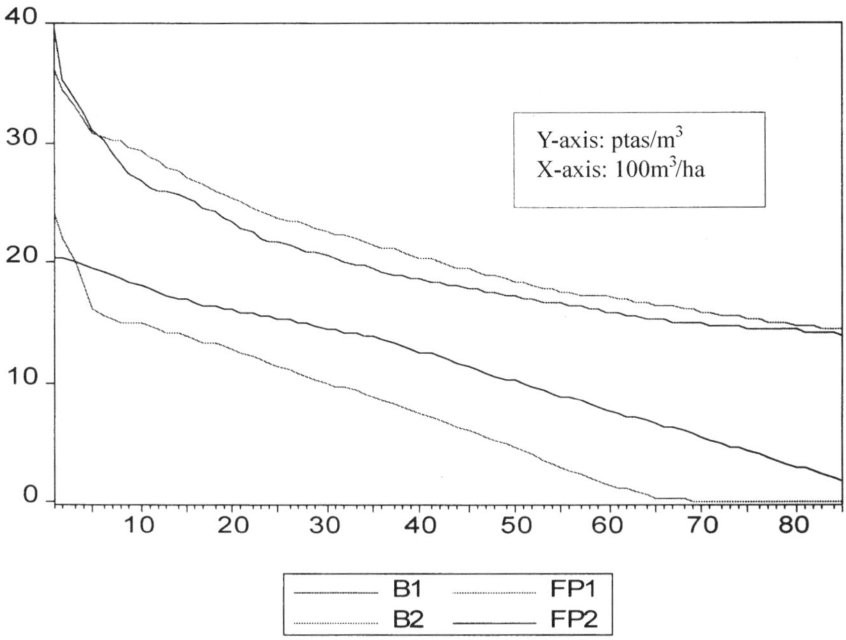

Figure 11.3. Inverse water demand functions for Bembezar and Fuente Palmera districts.

on growing cotton irrespective of any moderate change in water charges set forth by the Water Framework Directive.

Table 11.5 reports the regression results for the equation with acreage devoted to COP crops against the rest of the explanatory variables. These crops are less water intensive than cotton, sugar beets or alternative crops. The influence of water charges (WATCHARGE) differs across districts: it is not statistically significant for Bembezar, but it is positive and highly significant for the rest, specially for the Fuente Palmera and Genil Cabra districts. This implies that modern districts such as those just mentioned are more sensitive to changes in water charges because the soils are usually worse and the agronomic conditions more limited. Common results for both districts are the negative influence of the price of cotton, the Agenda 2000 scenario and the amount of water available.

As stated above, the mathematical programming model used to generate the results for the farms is based on an agronomic simulator called EPIC (Williams *et al.*, 1990)[1] which provides crop yields as well as values for nitrate percolation. A final regression was run to identify the factors that have an influence on this variable and the results are reported in Table 11.6.

The results show that the level of nitrate pollution is positively related to the amount of water available for irrigation. Higher water charges, the Agenda 2000 scenario and higher set-aside requirements would tend to lower the nitrate pollution. This means that Agenda 2000 combined with the Water Framework

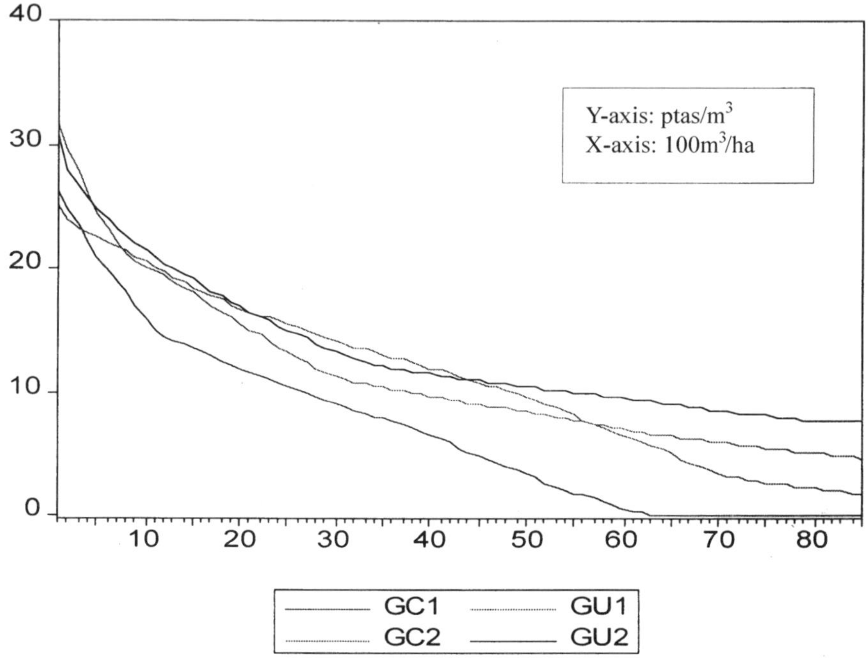

Figure 11.4. Inverse water demand functions for Genil Cabra and Guadalmellato districts.

Table 11.3.

Allotment situation	**Scenario 1** $P_w = 3$ ptas/m^3 1992 CAP reform	**Scenario 2** $P_w = 3$ ptas/m^3 Agenda 2000
Q_1 (normal)	Seller: GU Likely Seller: FP Likely Buyer: GC Buyer: BE	Seller: GU Likely Seller: FP, GC Buyer: BE
Q_2 (30% below normal)	Seller: GU Likely Seller: FP Likely Buyer: GC Buyer: BE	Seller: GU Likely Seller: FP Likely Buyer: GC Buyer: BE

FP: Fuente Palmera; BE: Bembezar; GU: Guadalmellato; GC: Genil Cabra.

Directive may have positive environmental effects. However, these results are only general trends. In some cases, signs of the effect of the Agenda 2000 scenario were found to vary with the level of water pricing, as can be seen in Figure 11.5. For the modern districts (Fuente Palmera and Genil Cabra) the effect of the Agenda 2000 changes from positive to negative for a higher water price. The influence of the level of support to the cotton industry is ambiguous;

Table 11.4. Regression of acreage devoted to cotton against other variables

Coefficients	Irrigation district Bembezar (ID-B)	Irrigation district Fuente Palmera (ED-FP)	Irrigation district Guadalmellato (ID-GU)	Irrigation district Genil Cabra (ID-GC)
INTERCEPT	−2.86 (−23.27)	−5.12 (−14.81)	−4.63 (−27.45)	−7.56 (−18.83)
WAT	0.00622 (382.28)	0.00953 (208.69)	0.00469 (210.17)	0.0089 (167.71)
WATCHARGE	−0.0025 (−0.23)*	−0.47 (−15.48)	−0.12 (−8.24)	−0.60 (−17.11)
AG2000	1.25 (15.7)	7.72 (34.45)	3.37 (30.84)	9.40 (36.1)
SETASIDE	−0.001629 (−0.23)*	−0.0176 (−0.88)*	0.0248 (2.53)	0.0101 (0.42)*
COTPRICE	9.61 (120.51)	19.23 (85.77)	15.07 (137.59)	22.78 (87.48)
n. of obs.	5504	5504	5504	5504
Adjusted R^2	0.9673	0.9059	0.9218	0.8729

(*t*-ratio in brackets). (*) not significant.

Table 11.5. Regression of acreage devoted to COP crops against other variables

Coefficients	Irrigation district Bembezar (ID-B)	Irrigation district Fuente Palmera (ED-FP)	Irrigation district Guadalmellato (ID-GU)	Irrigation district Genil Cabra (ID-GC)
INTERCEPT	64.88 (463.58)	83.53 (284.2)	97.78 (577.02)	92.27 (255.6)
WAT	−0.0030 (−163.85)	−0.0063 (−163.36)	−0.0034 (−154.58)	−0.0065 (−138.12)
WATCHARGE	−0.0032 (−0.26)*	0.2768 (10.75)	0.105 (7.077)	0.5462 (17.27)
AG2000	−6.82 (−75.17)	−11.73 (−61.60)	−7.09 (−64.57)	−10.45 (−44.67)
SETASIDE	−0.0266 (−3.28)	−0.1554 (−9.12)	−0.555 (−56.57)	−0.4054 (−19.37)
COTPRICE	−5.69 (−62.80)	−14.86 (−77.99)	−13.56 (−123.45)	−20.06 (−85.73)
n: of obs.	5504	5504	5504	5504
Adjusted R^2	0.8701	0.8711	0.8954	0.8425

(*t*-ratio in brackets). (*) not significant.

Table 11.6. Regression of the level of nitrate percolation (in kg/ha)

Coefficients	Irrigation district Bembezar (ID-B)	Irrigation district Fuente Palmera (ID-FP)	Irrigation district Guadalmellato (ID-GU)	Irrigation district Genil Cabra (ID-GC)
INTERCEPT	39.85 (513.77)	44.53 (162.37)	53.66 (392.05)	47.98 (170.92)
WAT	0.0057 (561.26)	0.0083 (228.75)	0.0046 (257.81)	0.0077 (208.54)
WATCHARGE	−0.02 (−3.04)	−0.57 (−23.73)	−0.317 (−26.48)	−0.60 (−24.35)
AG2000	−3.21 (−63.84)	−0.92 (−5.19)	−0.26 (−2.95)	2.38 (13.1)
SETASIDE	−0.017 (−3.79)	−0.110 (−6.94)	−0.306 (−38.61)	−0.243 (−14.93)
COTPRICE	−0.93 (−18.51)	3.83 (21.56)	3.50 (39.45)	4.01 (22.05)
n. of obs.	5504	5504		5504
Adjusted R^2	0.9833	0.9076	0.9281	0.8920

(*t*-ratio in brackets). (*) not significant.

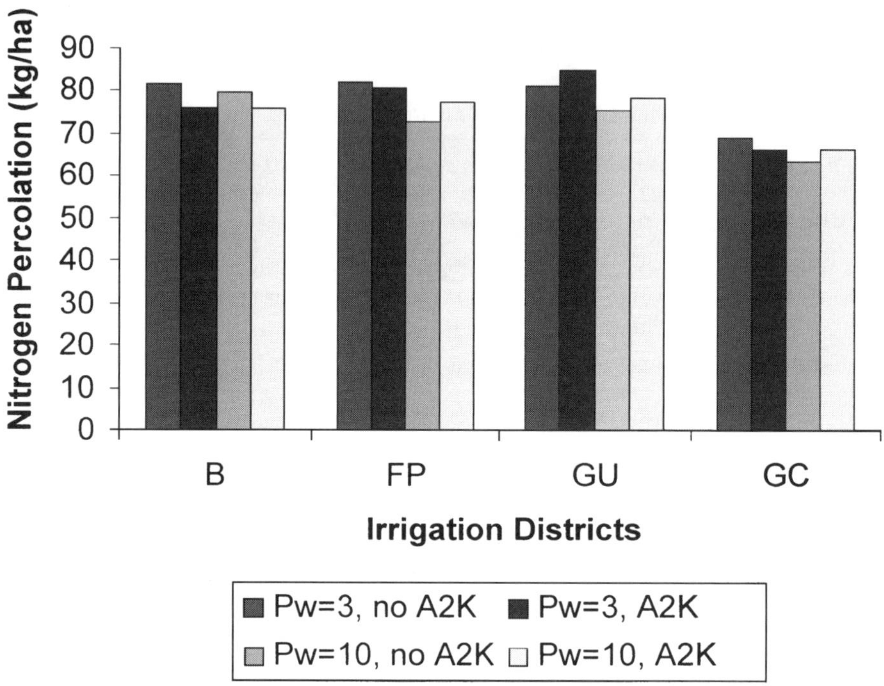

Figure 11.5. Simulated levels of nitrogen percolation for the four districts (normal water allotment considered for each one) under four different price and subsidy scenarios.

in Bembezar the current level of support tends to lower the level of nitrate percolation, whereas in the rest it is the opposite. The ambiguity of the cotton price effect is due to the various alternative cropping patterns that may replace those commonly developed, which are centred on cotton.

11.6. Final remarks

Farmers' water demand is largely dependent on commodity markets and the level and kinds of support delivered to various crops or agricultural activities. This paper contributes to the literature on irrigation water demand functions and attempts to evaluate the effects of two forthcoming European policies, Agenda 2000 and the Water Framework Directive, and one national policy, the reform of the water law, which will provide opportunities for water market transactions.

The combination of simulation results with regression analysis shows a few unambiguous effects for important variables within water management. For instance, Agenda 2000 sets forth important changes in the nature of the European crop support mechanisms. The result for our models show that the irrigators' shadow price for water will increase, although the Water Framework Directive will have an opposite effect. Whether the rises in the water charges will reduce the shadow price of water to a larger extent than the Agenda 2000 increases is not possible to ascertain from our results. In fact, our findings show that the relative magnitude of both effects will depend on the structural factors and natural endowments prevailing in each situation.

Another unambiguous result is that both European policies are expected to contribute to a reduction in the amount of nitrate pollution. Hence, in addition to the social gains accruable as a result of a better financing system for the water supply systems, the environment will be improved. To date, the literature on irrigation water demand functions has shown that gains to the farm sector of higher water charges are, at most, dubious. In fact, authors such as Sumpsi *et al.* (1998), Tsur and Dinar (1997) or Gardner and Warner (1994) contend that higher water charges for farmers would not deliver the benefits expected from them. This paper provides evidence that farmers would not respond to higher charges by changing their cropping patterns or reducing their water consumption to any significant extent. While this would mean that farm employment, food production or farm input purchases would not be significantly altered, farmers' quasi-rents could be lowered as farmers' benefits go down. This assertion is based on the assumption that water rights are appurtenant to agricultural land. However, the reform of the Spanish water law already passed by legislative means makes water marketing a legal activity. Although the likely results of the water markets are still unknown, the analysis must now be centred on the shadow price or dual value of water used in the agricultural sector.

Since the actual market cost of water will eventually become evident once water exchanges take place, farmers face two complementary types of incentives. One is due to the fact that water will be charged using volumetric mechanisms, and the other is that unused water will have a price in the market. Sales of redundant water will help farmers pay for the resources that will be used on the farm. These two effects will be quite significant.

Notes

1. EPIC has been calibrated to the region's soils and climate with funds from an EU research project (Contract No. 8001-CT91-0306).

References

Beare, S.C., Bell, R., Fisher, B.S., 1998, Determining the value of water: the role of risk, infrastructure constraints, and ownership, American Journal of Agricultural Economics, 80, 916–940.

Brill, E., Hochman, E., Zilberman, D., 1997, Allocating and pricing at the water district level, American Journal of Agricultural Economics, 79, 952–963.

Calatrava, J., Garrido, A., 1999, Examining effects from water trading on farm benefits, nitrogen pollution and labour market, Unpublished paper. Department of Agricultural Economics, Universidad Politécnica de Madrid.
Cummings, R.C., Nercessiantz, V., 1992, The use of water markets as a means for enhancing water use efficiency in irrigation: case studies in Mexico and the United States, Natural Resources Journal, 32, 731–755.
Dinar, A., Subramanian, A., 1997, Water Pricing Experiences. An International Perspective, World Bank Technical Paper No. 386. Washington, DC.
European Commission, 1997, Commission Proposal for a Council Directive Establishing a Framework for Community Action in the Field of Water Policy, COM(97)49. Brussels.
Fundenberg, D., Tirole, J., 1993, Game Theory, Cambridge MA: MIT Press.
Gardner, B.D., 1997, Some implications of federal grazing, timber, irrigation, and recreation subsidies, Choices, Third Quarter, 9–14.
Gardner, B.D.,. Warner, J.E., 1994, The central valley project water improvement act and water markets. Two steps forward – one step back, Choices, First Quarter, 4–9.
Garrido, A., 1995, La Economía del Agua: Análisis de la Asignación de Recursos mediante el Establecimiento de Mercados de Derechos del Agua en el Valle del Guadalquivir, PhD Dissertation. ETSIA. Madrid.
Garrido, A., 1998, An economic analysis of water markets within the Spanish agricultural sector: can they provide substantial benefits? in Easter, K.W., Rosegrant, M.W., Dinar, A., eds., Markets for Water – Potential and Performance, New York: Kluwer Academic Publishers, pp. 223–239.
Garrido, A., 2000, A mathematical programming model applied to the study of Spanish water markets, Annals of Operations Research, 94, 105–123.
Garrido, A., Mesquida, G., 1998, The cotton and tobacco sectors in the EU, in Tracy, M., ed., CAP Reform: The Southern Products, APS, Agricultural Policy Studies, Brussels, pp. 147–154.
Garrido, A., Varela-Ortega, C., Sumpsi, J.M., 1997, The interaction of agricultural pricing policies and water districts' modernisation programs: a question with unexpected answers, Paper presented to the Eighth Annual Conference of the European Association of Environmental and Resource Economists, Tilburg, the Netherlands, June 26–28.
Huffaker, R.G., Whittlesey, N.K., 1995, Agricultural water conservation legislation. Will it save water?, Choices, 4, 24–28.
Huffaker, R.G., Whittlesey, N.K., Michelsen, A., Taylor, R., McGuckin, T., 1998, Evaluating the effectiveness of conservation water-pricing programs, Journal of Agricultural and Resource economics, 23, 12–19.
Iglesias, E., Garrido, A., Sumpsi, J.M., Varela-Ortega, C., 1998, Water demand elasticity: implications for water management and water pricing policies, Paper presented to the World Congress of Environmental and Resource Economists, June 25–27, Venice.
Montginoul, M., Rieu, T., 1996a, Instruments de gestion de l'eau d'irrigation en France: exemple de la Charente, Ingénieries–EAT, 8, 3–12.
Montginoul, M., Rieu, T., 1996b, Instruments économiques et gestion de l'eau d'irrigation en France, La Houille Blanche, 8, 47–54.
Moore, M.R., Gollehon, N.R.,. Carey, M.B., 1994, Multicrop production decisions in Western irrigated agriculture: the role of water price, American Journal of Agricultural Economics, 76, 859–874.
OECD, 1999, The Price of Water. Trends in OEDC Countries, Paris: OECD.
Rainelli, P., Vermersch, D., 1998, Irrigation in France: Current Situation and Reasons for its Development, Unpublished manuscript from a study submitted to the OECD Environment Directorate.
Schaible, G.D., 1997, Water conservation policy analysis: an interregional, multi-output, primal-dual optimization approach, American Journal of Agricultural Economics, 79, 163–177.
Sumpsi, J.M., Garrido, A., Blanco, M., Varela-Ortega, C., Iglesias, E., 1998, Economía y política de gestión del agua en la agricultura, Secretaría General Técnica. Ministerio de Agricultura, Pesca y Alimentación (MAPA), Ediciones Mundi-Prensa, Madrid.

Tsur, Y., Dinar, A., 1997, The relative efficiency and implementation costs of alternative methods for pricing irrigation water, The World Bank Economic Review, 11, 243–62.

Varela-Ortega, C., Sumpsi, J.M., Garrido, A., Blanco, M., Iglesias, E., 1998, Water pricing policies, public decision making and farmers' response: implications for water polic, Agricultural Economics, 19, 193–202.

Williams, J.R., Dyke, P.T.,. Fuchs, W.W., Benson, V.W., Rice, O.W., Taylor, E.D., 1990, EPIC: Erosion Productivity Impact Calculator, Temple, Texas: United States Department of Agriculture

CHAPTER 12

Socio-economic and Institutional Factors Affecting Water Resources Management in a CAP Framework

Felisa Ceña and Dionisio Ortiz*

12.1. Introduction

The evolution of agriculture in arid and semi-arid environments has a traditional connection with water management. In Mediterranean countries, people are especially concerned about finding, storing, distributing and using water resources. This concern has not only been a question of hydraulic infrastructures or public works, but also a matter of rules, norms, and customs, the issue of who owns the water, how it should be used, and which practices are forbidden.

This chapter tries to show how some socio-economic and institutional factors play a relevant role in the management of groundwater water resources in southern Spain. The main argument is as follows (see Figure 12.1): on the one hand, agricultural activities use groundwater both as a source of water for irrigation, with overexploitation as a negative aspect, and as a sewer for input excess (i.e. pollution). On the other hand, groundwater management is conditioned by physical and environmental constraints, the latter having become crystallized into new rules and norms stemming from political initiatives.

But the way in which both of these factors affect each other is not independent of the institutional framework, which consists of water laws and customs, and the economic elements introduced by the Common Agricultural Policy (CAP) measures. Indeed, the importance of these factors means that it is not possible to try to modify the relationship between agriculture and groundwater resources without considering how economic signals from CAP and how the inertia of water institutions, both formal and informal, reshape this linkage.

The region chosen, Andalusia, is a good example of a semi-arid area where environmental constraints, water institutions and agricultural policy factors affect groundwater use in agriculture, which is the principal consumer (80% of the water resources available). Together, those three elements determine the situation, the problems, and the possible future trends of groundwater management in irrigated areas.

The relevance of irrigation areas in Andalusia can be easily estimated from

C. Dosi (ed.), Agricultural Use of Groundwater, 259–274.

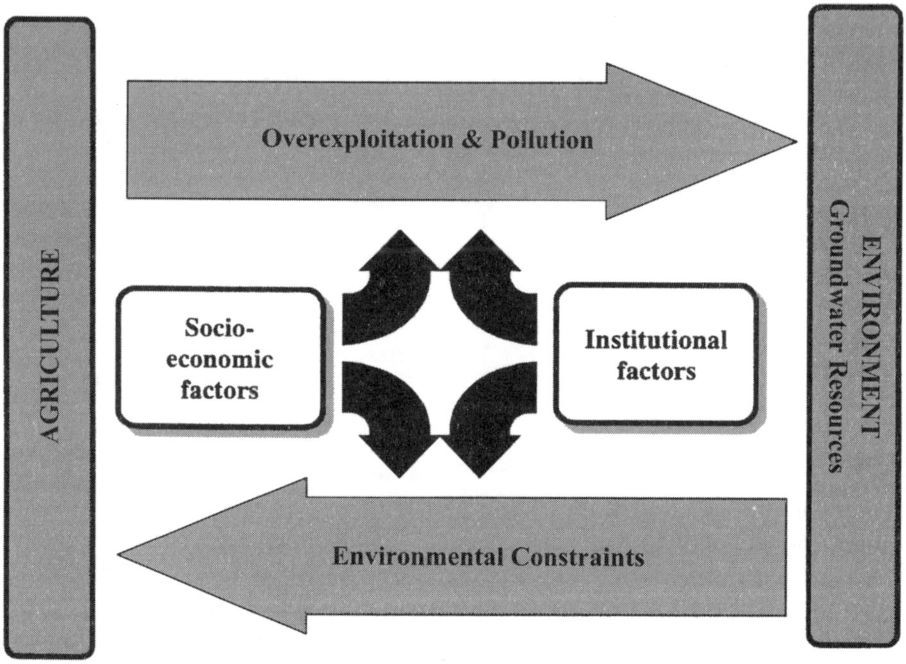

Figure 12.1 The role played by socio-economic and institutional factors in linking agriculture and groundwater.

Table 12.1. Agricultural water use in Andalusia

Water sources	Irrigated surface (1000 ha) Inland basins	Coastal basins	Total Andalusia	Estimated consumption (hm^3/year)
Surface water	516.3	91.0	607.3	2948
Groundwater	126.4	81.9	208.3	908
Total	642.7	172.9	815.6	3856

Source: DAP (1999) Inventory and characterization of irrigation in Andalusia.

available data: 16% of total agricultural area in the region is under irrigation. This area contributes 53% of the total agricultural output, and provides 60% of agricultural employment, especially important in a region with serious unemployment problems. The relative weight of groundwater irrigation is shown in Table 12.1.

As mentioned above, it is possible to distinguish two groups of factors affecting water management in the agriculture of southern Spain. On the one hand, there are institutional factors, rules and norms related to the legal aspects of water use. In this context, the two most important institutional elements conditioning agricultural water utilization are the property rights structure, and the users' contribution to water cost payment. On the other hand, socio-economic factors such as the role of agriculture as food supplier, the traditional relevance of this sector in national economies, and the fact of being the most important, even the sole activity in most rural areas, explains a long tradition of public intervention. Government protection has led to the development of some political instruments which have had a strong influence on agricultural activity, and therefore on the management of productive factors such as water. This protection has been both direct, in terms of price support and direct aids, and indirect, by easing the access to production factors, or providing information and formative aid to farmers.

In the European context, these socio-economic factors have been transferred to the CAP philosophy, the political framework controlling the main aspects of agricultural activity. CAP has therefore become a sort of transmission mechanism transferring social objectives and economic constraints to the norms regulating European agriculture. Consequently, an analysis of CAP measures could be utilized to give an approximation of the implications of these socio-economic aspects for agricultural water use.

The chapter has the following structure. First, both types of factors are presented and a brief justification of their role in water management is provided. Second, we focus on the effects of both types of factors, both separately and jointly, on the use of groundwater in agriculture and of the performance of administrative environmental constraints. Third, a brief discussion about the duality between security/flexibility of property rights is presented, as well as the implications of this duality in the evolution of irrigated areas. Finally, some conclusions and proposals are put forward.

12.2. Water policy: institutional factors

According to the Spanish Water Law of 1985, water is a commodity which is in the public domain: that is, the State owns the water and allocates it according to a set of priorities. Agricultural use is in a favoured place, preceded only by domestic use and environmental constraints. Once water is allocated to a specific user such as a farmer or an irrigation association, he or they have the right to use the flow that has been allocated. This right is conditional on appropriate management and is given for a specific piece of land (i.e. water cannot be used either on another farm or for a purpose other than irrigation). These rights are granted for a maximum of 75 years, but in the case of domestic use and irrigation, they can be extended if there has been no conflict with water planning authorities at the basin level.

However, property rights stemming from this administrative concession are more consolidated than law states. Indeed, policy instruments which could be used by water administration to adapt concessions to changing circumstances present serious constraints because of informal institutions. Farmers owning water property rights actually form an important lobby with enough power to confront administration. In other words, water administration bodies have the authority and the instruments to modify users' concessions and thus create new scenarios in which the water requirements are different to the way they are now, but experience shows that this possibility has not been utilized due to the political implications of creating such scenarios. This lack of flexibility can also be found in the procedure by which farmers obtain the right to use water. This is a complex and long administrative process which may last several years. The implications of this for agricultural water management will be analysed below.

The other institutional element which has heavy consequences for water management in irrigated land is the system by which users pay for the water. There is no price for water. In the most favourable circumstances, users of surface water only have to make a contribution to the hydraulic investments made by the State and operation and maintenance costs. This amount, calculated according to farm surface area, is independent of water consumption. There are thus no incentives to save water. On the other hand, users of groundwater afford energy costs for pumping water calculated on the volume extracted. Consequently, since the afforded costs depend on the volume extracted, for these users irrigation systems are more efficient.

In addition, agriculture is exempted from the consequences of another factor for which other water users pay: water pollution. There is a tariff for those users who divert water and then return it to the general system. This tariff is intended to reduce water pollution: it is an application of the polluter pays principle. But farmers are exempted from this duty. In its water policy, therefore, the administrative body does not acknowledge that there may be excessive water used on farms, nor is it concerned about agricultural non-point pollution.

12.3. Socio-economic aspects affecting agricultural water management

As suggested above, CAP indirectly affect agricultural use of water. The main element affecting input utilization (and therefore water resources management) is crop profitability. Profitability largely determines what resources will be used, how they are going to be managed, the intensity of use, and the timing of utilization.

The incentives introduced by subsidies (both coupled and decoupled from production levels) are thus very interesting to analyse in terms of their effects on water management in agricultural systems, since these are mainly responsible for political modifications of agricultural profitability. The following sections give an outline of both types of subsidies.

12.3.1. *Production subsidies*

The effect of production subsidies in European agriculture from the 1960s until the McSharry reform in 1992 is well known by politicians, economists and farmers. At the macro level, the rise in production promoted through artificially high prices exceeded all expectations and the original objective of assuring market stability turned into a deep financial problem for EU budgets. At the micro level, one of the main outputs of production subsidies is an intensification of the use of natural resources. The stress on land and water has generated large environmental problems that continue nowadays.

In the years following the CAP reform, many of these market supports started to change. It was a period of institutional price reductions and the birth of direct compensatory aids, whose implications for water use will be outlined later. Some direct subsidies for specific sectors such as the one applied by the CMO to olive oil also came into effect. In spite of the recent Commission's attempt to transform production subsidies into aid given according the number of olive trees, pressure (exerted mainly by the Spanish government) has led to the maintenance of this measure.

12.3.2. *Direct aids*

The so-called direct aids for COP crops (cereals, oil and protein seeds) were designed to compensate farmers for price reductions in European agricultural markets. This kind of subsidy was estimated according to a reference regional yield based on historical data. It failed to take an important point into account: the difference between dry and irrigated yields within the same region and for the same product. The discrepancy between yields varies according to product and regions, but it can reach differences of 65% between dry and irrigated regional yields in traditional irrigation districts in southern Spain.

Different levels of direct aid for water management have arisen because administrative recognition of irrigated land needs acknowledgement of water rights linked to that land. That is, illegally irrigated farms cannot claim that advantage because of the lack of water rights. Furthermore, to obtain the subsidies corresponding to irrigated land, it is not only necessary to be in a irrigation district, but also to use the water.

12.4. The effects of institutional and socio-economic aspects in the link between agriculture and groundwater management

How do the factors presented above affect agricultural activity, what are the consequences for groundwater resources, and how do environmental constraints related to groundwater condition such activity? How do these factors constrain and condition the future of irrigated agriculture (and therefore the future of water management in the sector) in a region like Andalusia? To answer these questions, two main effects will be described: on the one hand, the role of economic signals and the institutional structure on how agriculture affects groundwater resources (mainly over-exploitation and pollution); on the other, the way in which new policy tools, implemented to try to limit environmental damage to groundwater, have to face socio-economic and institutional obstacles which could limit, or even destry, their effectiveness.

12.4.1. *Environmental damage caused by agriculture: economic and institutional framework*

The main element here is how illegal use is promoted, use which often brings over-exploitation of groundwater and pollution problems. Illegal use in agriculture can be understood as being the abstraction of water, normally pumped from a river or from groundwater, to irrigate without the required administrative permission. This practice is very widespread: only half of all water usage in Spain is registered in the water register, a compulsory requisite introduced by the 1985 water law for every user. Reasons explaining the broad scale on which illegal usage is practised come from both the high profitability of specific crops (sometimes distorted by CAP subsidies), and the lack of flexibility of water institutions.

Two elements worsen this lack of administrative control. First, the difficulty involved in controlling individual wells in each farm, since these are sometimes hidden, or are permitted to have a different flow to others, or a different number of functioning hours. Only 23% of the estimated number of wells in Spain are registered, 6% in Andalusia.[1] Second, and with important implications for agriculture, the lack of political will to stop some of the processes that take place without administrative permission when new irrigation districts are started up.

These new irrigation areas are based on crops whose profitability has risen in the last years. A clear example can be found in Jaen, the most productive Spanish province for olive oil, where many traditional dry lands have been put under irrigation. This is an instructive example of how a rise in crop profitability due to political decisions is able to affect the management of natural resources in agricultural systems. Application of the olive oil CMO has resulted in a subsidy which has increased the income received by farmers during the nineties by more than 30%. In addition, the difference in productivity in irrigated farms vis à vis non-irrigated ones is, on average, 60%. Both elements, together with the rise in market prices in recent years (because of a reduction in supply due to rain deficit) have made farmers race in search of water, whatever the source and whatever the difficulties.

The effect on the olive sector in Jaen of the convergence of the above factors is that 127 000 hectares of new irrigated land have been brought into production, 40 000 from groundwater resources (Corominas, 1999). This has been a bottom-up process where the administrative body has taken a passive role, merely witnessing farmers' initiatives. Further problems have when water planning authorities at the basin level have stated that only 50 000 ha are going to obtain a legal concession. The lack of public control could have other negative consequences: there are no studies about the potential effect of massive groundwater over-pumping on piezometric levels in years of rain deficit, of the use of urban waste water for irrigation, or of the external economies that could be damaged if those new water uses are eventually forbidden.

But while private initiatives are being fostered by artificial economic signals which are leading to some crops having a high profitability, the search for business opportunities is being obstructed by the rigidity of water institutions. Indeed, as mentioned above, the bureaucratic procedures needed to obtain a concession involve long and complex processes which can take several years to complete, with files often lying around forgotten because of the inadequacy of administrative means, failure to make decisions of a technical nature, and, especially, a lack of political will. As a result, the transaction costs involved in obtaining water property rights (in terms of lost opportunities) are so high that users usually start abstracting water at the same time they put in a request for administrative permission, in spite of the economic sanctions they face. Since users start abstracting water before technical studies about groundwater availability or damage to third parties are carried out, this leads to further loss of administrative control.

Nowadays, most of the new requests for water concessions are in the final stage, awaiting the construction of new barrages, some of which have been put on hold because of environmental conflicts about the impact of these infrastructures. The EU, which has imposed various environmental constraints on these infrastructures, is now becoming aware of these conflicts.

Another aspect of administrative failure to control illegal usage is the lack of information about the potential damage to the environment, for instance, impacts due to non-point source pollution. High pollution levels occur, and in coastal areas, overpumping has also lead to the intrusion of saline water from the sea. Water stress, both from qualitative and quantitative perspectives, is now becoming evident in these areas. Demographic projections foresee a larger concentration of population and activities along the Spanish Mediterranean coastal fringe. But there are not only problems along the coast. Indeed, in the Guadalquivir basin – the main river basin in Andalusia – there are 20 points at which, according to the Ministry of the Environment, water quality has a GQI,[2] catalogued as inadmissible. Some of these situations are due to industrial or urban waste water, but others may well be related to agricultural pollutants.

12.4.2. *Environmental constraints versus economic and institutional factors*

The way in which environmental constraints condition the use of water is, as has already been stated, related to both economic incentives and institutional structure. In the case of agriculture in Andalusia, these are exemptions from the polluter pays principle, and the weakness of definition of property rights.

Exemption from the polluter pays principle

The protection of agriculture in the CAP context conflicts with the application of the polluter pays principle. It is not only a matter of it being difficult to measure agricultural non-point source pollution. It is also a matter of institutional design. As stated above, agriculture is not legally considered to be a polluting activity. While urban and industrial users have to pay an extra tariff in recognition of their environmental impact on water quality, regardless of whether there has been compulsory installation of purifying plants, agricultural non-point pollution has not been considered to be a source of environmental damage in economic terms. The reasons for this may be both the desire to avoid imposing more charges on farmers for political reasons, and technical difficulties involved in controlling and measuring waste production at the farm level. In any case, it is another example of the lack of policy incentives to reduce the environmental impact of agricultural activities.

It is difficult to measure agricultural pollution. The instruments used to reduce environmental damage are usually compulsory, and based on environmental standards, because other economic instruments present higher transaction costs and administrative difficulties (Weersink and Livernois, 1996). One example of the instruments used is the European Directive 91/676/CEE on protection of water against nitrate pollution due to agriculture. This first step towards recognition of the impact of agriculture on water quality failed to include economic tools for reaching the objective. Indeed, the Directive established an obligation for administrative bodies to undertake an analysis of sensitive areas and create guidelines for environmental friendly practices. These practices would be voluntary for farmers, but compulsory for those located in these areas.

A weak definition of property rights

In spite of water legislation, practices reveal a lack of definition of property rights. Indeed, in some situations, users can extract water without being subjected to controls and norms governing usage. The situation is closer to a free access system than to monitored management. Since new users can utilize water without the required administrative permission, and no institutions restrict this illegal usage, environmental problems stemming from free access are made possible.

This question is closely connected to how externalities and property rights are defined, and the relationship between the two. The Coase theorem is well known from economic textbooks: the impact of an activity over another one is considered as an externality depending on the allocation of property rights. However, in a context where property rights, specially those related to groundwater resources, are not clearly allocated, illegal usage, lack of control, and difficulty in defining different positions in relation to environmental externalities arise.

Another element contributes to making difficult the evaluation and the internalization of externalities. While 78.5% of land irrigated with surface water is managed in a collective way, only 37.7% of irrigated land using groundwater resources is managed under the same type of system. In other words, for most agricultural groundwater usage, it is very difficult to introduce policy tools that can take account of interdependent individual usage. This makes it unlikely that implementing policy tools in surface and groundwater irrigation zones will have uniform success.

Although the Water Law gives basin administrative authorities sufficient power to intervene when both over-pumping and pollution occur, in practice, the lack of monitoring and of information about aquifers makes intervention and control of users difficult.

12.5. The difficult balance between security and flexibility in the structure of property rights

Within the context of property rights, a topic which has clear implications for water management in agriculture is the balance between the security and flexibility of those rights.

In Andalusian agriculture, there is a conflict between structural adjustment taking place mainly through technological innovation, and the structure of property rights. The effects of this conflict can be found in two types of agricultural irrigation systems: traditional irrigated areas and emerging agricultural systems.

Traditional irrigated areas

On the one hand, there are traditional irrigated areas with low profit production largely maintained by European subsidies (mainly arable crop such as cereals, cotton, sunflower, and sugar beets). For historical reasons, these irrigation districts have consolidated water property rights. These were obtained by means of a top-down process in which the farmers did not have to do anything to obtain the water rights, with the hydraulic infrastructures allowing irrigation. Thanks to fortunate conditions, farmers went almost overnight from having dry land to being able to improve their yields and hence their income.

The fact that these farms are now irrigated provides an extra bonus: since the 1992 CAP Reform, irrigated farms have received larger compensatory payments than dry farms. Because of exogenous political decisions, they have been able to operate in exactly the same way that they have operated for several years, but with very different results. This has lead to a strengthening of their institutional position, as obtaining direct aid for irrigated land requires having the right to irrigate. Farmers in these areas have chiefly been concerned with trying to protect their position in the water rights structure. These farmers have become an active lobby. At present, they are expanding their power basis to include other social actors (labour unions, local administration, and agricultural organizations) whose main objective is the creation of supply policies allowing them to maintain their water rights.

The reasons given by farmers in these traditional areas to defend their property rights are historical and social, and they use them to exert pressure on administrative authorities to put public investment into creating new barrages and setting up modernization projects. But the modernization of traditional irrigated areas has become a hotly debated issue. Mainly because of surface irrigation systems, the technical efficiency of some irrigation districts is no greater than 50–60%: i.e. almost a half of the water taken for irrigation does not reach the targeted crops. In many districts with semi-arid conditions, this is leading to proposals for ambitious modernization projects which need both big public expenditure and investments by farmers. However, the appropriateness of these initiatives is being questioned. The two arguments against them are the following: (i) on the one hand, it is well known that water losses are picked up downstream due to the basin structure, so water savings mean a reduction of groundwater flows and their availability to other areas downstream; and (ii) on the other hand, the uncertain future of traditional agricultural systems within the changing context of agricultural policies could bring the medium and long-term profitability of them into doubt, both from the private and the public point of view (Ortiz and Ceña, 1997).

Emerging agricultural systems

On the other hand, there are emerging agricultural systems well integrated into national and international markets (though some of them receive European aid, such as producers of olive oil) which are trying to compete in the use of water resources. Since the water law admits a kind of prior appropriation principle, establishing that every new concession should be required to avoid damage to other holders of rights, conflicts arise.

The extreme within this group is irrigation in coastal areas. These agricultural systems are characterised by intensive utilization of water (mainly groundwater) with a fragile property rights framework. They are also in strong competition with other water uses (tourism). The problems brought about by this kind of water usage include some environmental damage caused by intensive use of both water and agro-chemicals.

This combination of institutional and environmental factors has put water administrators into a difficult position in relation to the farmers, who are willing to pay the real cost of water supply and infrastructure investment in a bottom-up process, and in relation to legal and environmental constraints which would limit the development of an activity that has meant economic growth and creation of employment.

These new irrigation districts are usually more competitive in agricultural markets, especially in products with more added value. In addition, they are also more efficient from a water management point of view. Farmers in these zones, where the property right situation restricts access to the legal consolidation of their water usage, are therefore critical of the use of large amounts of water in traditional irrigation districts.

In short, institutions governing water rights do not adapt easily to changing circumstances. While the most consolidated water rights belong to irrigated areas whose agricultural productions are in a deep crisis within the CAP, new agricultural areas have to face an unstable situation. In other words, in the property rights structure, it is difficult to find a balance between having security and maintaining flexibility.

Both types of property rights have important implications for surface and groundwater management. The desire for security means that owners will include long-term costs in their production planning. From the groundwater perspective, this would suggest that the rate exploitation/recharge will be respected. However, as only 37.7% of irrigated land using groundwater resources is managed in a collective way, the problem is likely to remain. Most of the groundwater resources are used under conditions of free access. Consequently, users do not have any incentives to exploit them in a sustainable way.

The need for flexibility means that new users willing to pay the real cost of water, or at least to invest in irrigation systems without public aid, would need to have access to water rights in a monitored way. Institutions adapt to new circumstances and demands, trying to answer to changes in social goals. Since it has been demonstrated that water rights within this context are too rigid, these new demands are leading for mechanisms to obtain water in a way that is parallel to having legal control. In other words, the lack of institutional flexibility is also inducing practices which could lead to environmental damage to groundwater resources.

A point of equilibrium could be obtained by making property rights secure enough to induce users to assume long-term costs (managing groundwater resources in a sustainable way). To reach this objective, collective and integrated management should also be promoted. But property rights also have to be flexible enough to allow adaptations to new social goals and emerging demands, giving clear signals to users about the evolution of property rights structure.

Table 12.2 summarizes the views presented above. It describes the situations and prospects of different agricultural systems in terms of productive capability and some aspects of how their water rights are structured.

12.6. Final remarks

It has been shown that both types of factors, water institutions and CAP signals, condition the bi-directional linkage between agriculture and groundwater resources. On the one hand, the property rights structure is able to

Table 12.2. Characterization of irrigated agricultural systems in Andalusia

Agricultural systems	Extensive systems	Semi-intensive systems	Intensive systems
Brief description	Traditional areas with obsclete irrigation systems (arable crops)	Areas in a process of modernization, both from a technical and an agronomic perspective (trees and horticulture)	Coastal zones, with greenhouses and other intensive systems (intensive horticulture)
Water rights structure			
Water rights position	Consolidated and old water rights with a strong lobby defending their interests	Without strong water rights, but with the administrative support	Problems in consolidating their water use in a framework of strong competition for resources (tourist stress)
Environmental impact	Inefficient use of water. The irrigation system produces soil losses and agro-chemical dragging	Unplanned use of water resources (surface and groundwater, and urban waste water)	Intensive and uncontrolled use of groundwater (pollution and sea water intrusion)
Main demands	Maintenance of the prevalent rights position. Modernization of irrigated areas with public funds	Recognition and consolidation of their position	An increase in water resources through transfers from interior river basins

distort the effects of environmental policy instruments, since it defines the balance between flexibility, the adaptability of water use and management to changing circumstances (market liberalization, price fluctuations, new demands), and security, allowing individual interests to coincide with social ones from a management perspective. On the other hand, artificial profitability due to CAP subsidies fosters intensive use of water resources, exceeding in some cases the recharge rate, while the water law has forgotten that agriculture is also a polluter.

In respect of the manner in which the property rights structure affects the agricultural use of groundwater resources, three main topics seem to be the most relevant:

- The failure to define property rights is aggravating the problems of groundwater management.
- The high transaction costs involved in getting new water rights are inducing illegal usage.
- There is growing conflict between farmers in different irrigation systems stemming from the rigidity of property rights. This conflict has both technical and institutional aspects.

But there are not only institutional factors affecting this relationship. Economic signals introduced by CAP also influence farmers' behaviours and strategies, especially those relating to rates of resource utilization. Production subsidies for olive groves is leading to the transformation of dry farms into irrigated ones, frequently using groundwater. Other direct public aid, initially uncoupled from production levels, also impedes more rational water management. It has become a factor reinforcing the property rights status quo, thus limiting the flexibility of rights.

All those factors also condition the effectiveness of environmental regulation. In regard to this, three factors stand out as being important ones:

- Administrative failure in the monitoring of groundwater use is limiting the introduction of environmental instruments which would bring about an improvement in natural resources management. The lack of information about agricultural utilization of groundwater, and its real effects forms a serious obstacle for policy tools, which need a thorough knowledge of these elements in order to be effective in any real sense.
- In addition, the individual character of groundwater exploitation, with its low rate of collective management, means that external dis-economies have little effect on the decision-making processes. That is, those resources are likely to experience environmental damage as a consequence of free access conditions.
- Finally, the willingness to protect agriculture against economic menaces has meant the lack of application of the polluter pays principle. This protection hinders the use of economic instruments, especially those based on economic

sanctions. European efforts to include environmental goals in agricultural policy are based more on positive incentives schemes (even to address negative environmental externalities) than on the use of sanctions, although economic theory suggests that these would be more correct.

Some proposals for actions to redress the above will now be presented. The first focuses on redefining the relationship between water and agricultural policy. This linkage should be reconsidered at two levels: at the planning level and at the instrumental level. Within the planning framework, water management in Spain is related to national planning which considers demand and supply projections, and establishes priorities for use which depend on the condition of hydrographic basins. Within this context, long-term water demand from irrigated areas should be consistent with the future of agriculture within the framework of the EU. Those planned demands must be compatible with the foreseeable decrease in the prices of most agricultural commodities in line with world prices, a trend started with the MacSharry reform and confirmed in Agenda 2000. In that future scenario, it seems likely that there will be a reduction of the surface area devoted to traditional arable crops (cereal and oilseeds). This might jeopardize large irrigation projects carried out in the past and even some of the present modernization projects, since these are solely technical infrastructure improvements which do not take agricultural market considerations into account. As both kinds of irrigation projects are mainly financed by public budgets, future water planning ought to realize that new scenarios resulting from the WTOs Millennium Round may have to be considered. But this redefinition at the level of intersection of both policies should also be applied at the instrumental level, i.e. through the introduction of policy tools aiming at both agricultural and water goals. This is not a new approach since it is a sort of so-called cross-compliance, a term often used in relation to the new agri-environmental proposals in the EU. Cross-compliance means the conditioning of subsidies to the fulfilment of some environmental goals, friendly agricultural practices in natural systems, that can be linked to the imposition of some constraints in order to reduce water problems (pollution or overpumping). Agri-environmental measures, timidly introduced by the CAP as accompanying measures, are being adopted in specific areas, but in a way absolutely uncoupled to any water planning. General environmental objectives, but no kind of water deficit reduction or more rational use, are being pursued. Dual instruments pursuing both agricultural and water goals could therefore become a powerful means for integrating objectives and obtaining better results.

The second proposal has to do with the debate about the subsidiary character of water prices, a debate very actual in the European context.[3] In this discussion, positions defending either wider liberalization aimed at facing new global trends, more open markets, or the maintenance of public intervention are being taken up. However, the question of what effects further liberalization of water management will have on surface and groundwater resources still remains unanswered.

One of the effects could be a water transfer to non-agricultural use, where willingness to pay is higher. Several studies analysing the impact of these water transfers (usually from rural to urban areas) show how external costs imposed on rural economies exceed the private benefits to those who directly receive compensation, i.e. farmers (Charney and Woodard, 1990; Chan, 1990). This argument is closely related to the community value of water in rural areas. Values like justice, participation and local control, opportunity, and caring for the resources are historically linked to water availability and management in rural zones (Brown and Ingram, 1992). The cost of the loss of these values should also be considered when designing and implementing new water policies. That is, the new water policies must protect the social aspects of water resources, employment and wealth creation, rural development and maintenance of land use, and food security (Fereres and Ceña, 1998). Water, as an essential natural resource for human life and survival of agriculture in semi-arid rural areas, should be supplied by the state in conditions acceptable for those uses. The big questions are these: who is going to obtain subsidized water conditions, how are they going to get them, and under what circumstances?

A maturing water economy (Randall, 1981), like the Andalusian one, requires changes in the institutions governing it. This is the path that Spain has embarked upon through reforms in the water law which are aimed at introducing more flexibility in property rights by allowing water markets, even though this is being done under administrative control. Nevertheless, it is too early to know what the effects will be.

Notes

* The authors thank Dr. Fernando Andrada for valuable comments on preliminary versions of this paper.

1. Data from the Ministry of Environment (1999) *White Book of Water.*
2. The General Quality Index is a 0–100 index expressing 23 water quality parameters.
3. In 1987 the European Commission adopted a proposal for a *Water Framerork Directive* (COM(97)49), subsequently amended, and currently negotiated by the European Parliament and the Council of Ministers which, inter alia, is aimed at ensuring that, in all Member States, the price charged to water users is based on the full costs of water (full-cost recovery principle).

References

Brown, F.L., Ingram, H.M., 1992, El valor comunitario del agua: consecuencias para los pobres de las zonas rurales del Sudoeste, in Aguilera, F., ed., Economía del agua, Madrid: Secretaría General Técnica, MAPA, pp. 79–107.

Chan, A.H., 1990, Rural community values in groundwater marketing, Journal of Economic Issues, 24, 463–472.

Charney, A.H., Woodard, G.C., 1990, Socioeconomic impacts of water farming on rural areas of origin in Arizona, Journal of Agricultural Economics, December, 1193–1199.

Corominas, J., 1999, Papel de las aguas subterráneas en los regadíos, Jornadas sobre el papel de las aguas subterráneas en el Libro Blanco del Agua en España, Madrid.
DAP, 1999, Inventory and characterization of irrigation in Andalusia, Junta de Andalucia, Sevilla.
Fereres, E., Ceña, F., 1997, Social benefits and environmental constraints of irrigation in an era of water scarcity, in Water: economics, management and demand, International Commission on Irrigation & Drainage. Proceedings of the 18th European Regional Conference Water – an economic good, Oxford, UK, pp. 128–136.
Ministry of Environment, 1999, White Book of Water, Madrid.
Ortiz, D., Ceña, F., 1997, La modernisation des systemes d'irrigation: une innovation institutionelle?, Le cas de l'Andalousie (Espagne), Séminaire International: Agriculture et Développement Durable en Méditerranée, March 10–12.
Randall, A., 1981, Property entitlement and pricing policies for a maturing water economy, The Australian Journal of Agricultural Economics, 25, 195–220.
Swaney, J.A., 1988, Trading water: market extension, social improvement, or what?, Journal of Economic Issues, 22, 33–47.
Weersink, A., Livernois, J., 1996, The use of economic instruments to resolve water quality problems from agriculture, Canadian Journal of Agricultural Economics, 44, 345–353.
Whittlesey, Norman K. and Huffaker, Ray G., 1995, Water policy issues for the twenty-first century, American Journal of Agricultural Economics, 77, 1199–1203.

CHAPTER 13

The Environmental Impacts of CAP: An Overview of the Present State of Knowledge and Research Needs

Floor Brouwer

13.1. Introduction

Policy makers look to researchers to provide pertinent and sound evidence for problems and well-researched concepts to inform public decisions and choices. Science is expected to provide clarity and certainty. Recent trade negotiations provide a good example of how research can be influential. For example, the settlement on agriculture in the last GATT Round was influenced by research results and involvement and debate between European and American economists. Before new policy is negotiated at the EU level, there is a need to assess the impact of such policies, particularly at the Community and Member State level. In preparing proposals, the European Commission normally makes use of available research findings. A mismatch between the provision of scientific evidence and the eventual decision- making process might result in ill-informed courses of action (Hoogervorst, 1996).

Above all, it is critical that consensus be achieved if scientific results and theories are to play their part in the policy debate and in developing proposals to reform agricultural policy. Such consensus between economists certainly paved the way for agreement on proposals to reform agricultural policy. Such a consensus is apparent in the report from the group of experts charged with outlining the principles that are aimed at guiding CAP towards integration of environmental and rural development objectives (European Economy, 1997).

When the focus of CAP reform was essentially economic, achieving scientific consensus was more straightforward than it is at present, since there is now much less agreement on the environmental implications of policy reform. Opinions are more divided and there is a greater range of perspectives from various fields, perspectives that reflect the increasing complexity of the issues involved. In consequence, environmental linkages are less integrated into policy formulation than economic linkages. Dialogue between all parties, and particularly between policy makers and administration, is necessary to resolve the dilemma that has been created.

C. Dosi (ed.), Agricultural Use of Groundwater, 275–289.

Scientific agreement on the environmental implications of reforming agricultural policy remains limited. As well as this, it is still too early to assess the impact, as the Commission reported in its evaluation of the progress of the Fifth Environmental Action Programme, published in 1995.

This chapter provides an overview of the way that the environmental impacts of CAP are being viewed at present. In addition, some research gaps are identified and suggestions for further research are presented.

Some aspects of current farming practices will firstly be described, and the interaction between agricultural policy and the environment examined.

13.2. Agriculture and its interaction with the environment

Two dominant trends in current farming practices are intensification, concentration and specialization in some areas, and marginalization and abandonment in others (European Commission and Eurostat, 1999). They both involve a move away from traditional forms of low-input, labour-intensive cropping and livestock production, which have characterized most of Europe for many centuries.

Intensification and *specialization* involve the development of capital-intensive and geographically specialized farming of the sort that is mainly found in regions where agriculture is most productive. Some regions may have competitive advantages over others because of better biophysical conditions, more rationalized farm structures and the integration of primary production with food processing industries or through well-equipped farm extension services. Pig production, for example, is largely concentrated in regions which have an infrastructure facilitating production and processing, and having easy access to the main consumption regions or harbours for the import of material for the production of compound feed. It is primarily concentrated in Denmark, the Flanders region in Belgium, the Netherlands, Bretagne and the Po Valley area in Italy. Almost half of the pig population in EU-15 is currently grown on a small percentage of the holdings which have pigs. The numbers of holdings with limited numbers of pigs (less than 200) are also tending to decrease, while the pig population is increasing most rapidly in holdings which have more than 1000 animals (Table 13.1). Nitrogen pollution problems in Europe are highest in regions where agriculture has been specialized in intensive livestock production. It is essentially due to an excess amount of manure compared with the available land.

Marginalization and large-scale abandonment of agricultural land tends to occur in remote areas with unfavourable economic or social conditions, or on less fertile land where traditional extensive agriculture is threatened by its inability to compete effectively with intensive production in other regions. Abandonment, degradation and economic decline currently threaten the extreme north and south of Europe, where harsh natural conditions, poor soils

Table 13.1. Structure of pig farms in 1995, by herd size

Animals per holding	Holdings (× 1000)	Animals (× 1000)
1–2	611	832
3–9	228	1026
10–49	116	2794
50–99	41	2920
100–199	39	5563
200–399	38	11115
400–999	47	29927
⩾1000	27	56229
All classes	1146	110406

Source: Eurostat.

and remote locations increase the costs of agricultural production and rural populations are decreasing. In the southern part of Europe, marginalization and abandonment are significant problems across much of the interior of southern France, the Iberian Peninsula and Greece, and in many parts of Italy. In Spain, for example, the abandonment of marginal land with low productivity will potentially affect about 12 million ha of land, having a major impact in the form of soil erosion, fires, loss of biodiversity and landscape deterioration in general (Varela-Ortega and Sumpsi, 1998).

Figure 13.1 shows animal density per hectare of utilized agricultural area at a regional level. Animal density exceeds 2 LU/ha in the Netherlands, Belgium, Bretagne (France) and Lombardy (Italy). A stocking density of 2 LU/ha is considered to be close to the amounts of nitrogen from livestock manure which, according to the rules of the nitrates directive (as required by legislation in Germany, to name one country), is not allowed to be exceeded. A manure unit is considered to be close to 80 kg of nitrogen.

Intensification and specialization is leading to problems for landscape and bio-diversity, but also for water, soil and air. Problems related to marginalization and abandonment include the abandonment of water management systems, the incidence of soil erosion, increased risks of forest fires and major floods. This dual development of the two ongoing and opposing forces of intensification and marginalization and abandonment can be observed in the EU.

We will now describe some of the issues that are of major importance in the interaction between agriculture and water (quality and quantity). It is thought that nitrates in groundwater will remain at a stable level, and pesticides in groundwater will continue to remain a problem (EEA, 1999).

13.2.1. *Nitrate pollution problems from agriculture*

Contamination of groundwater and surface waters caused by high levels of production and use of manure and chemical fertilisers is a serious problem in

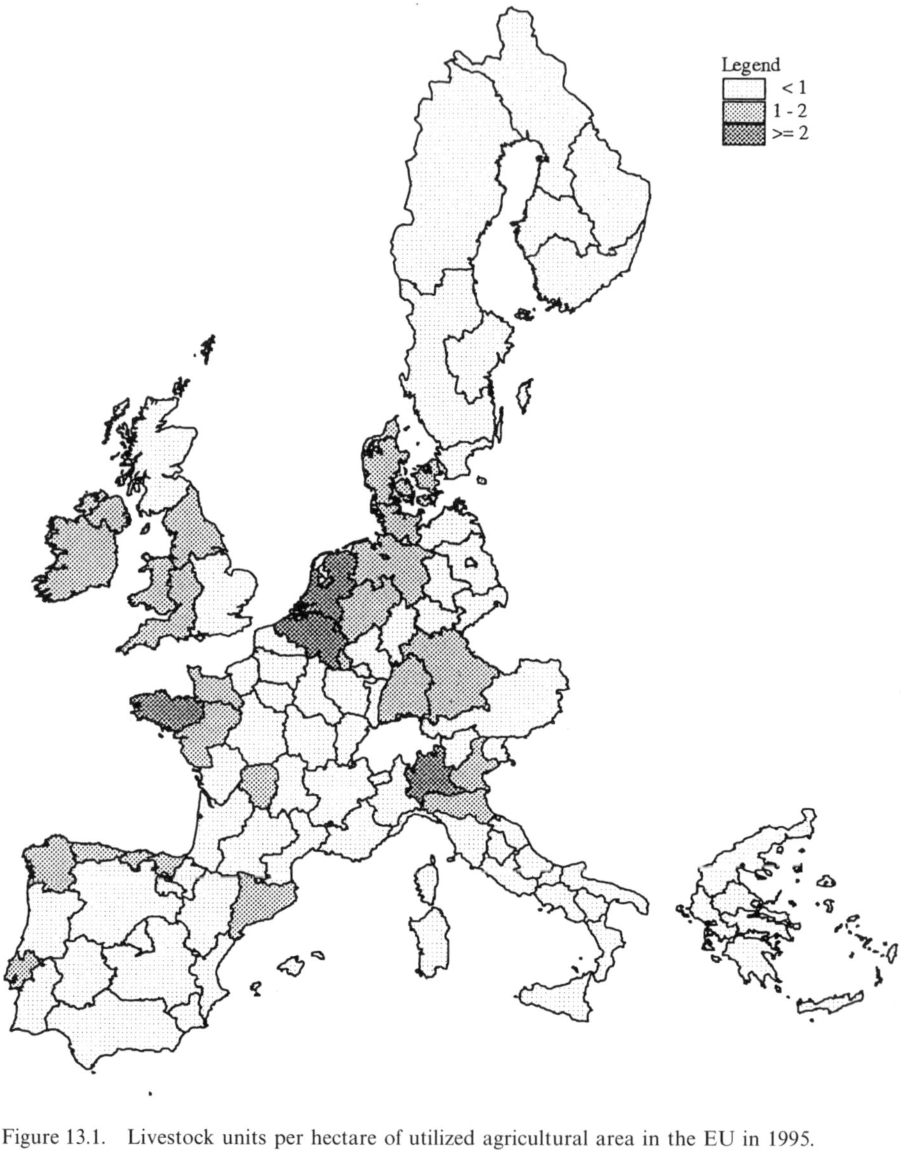

Figure 13.1. Livestock units per hectare of utilized agricultural area in the EU in 1995.
Source: Eurostat (Farm Structure Survey); adaptation LEI.

some parts of Europe. The maximum admissible concentration of nitrates for drinking water (50 mg/l, following the standards formulated by the World Health Organization) are exceeded on about 20% of the agricultural land (Stanners & Bourdeau, 1995). Problems are most acute particularly in regions where there are concentrations of intensive livestock production (mainly pigs and poultry) or large areas of specialized crop farms (including intensive horticulture). The supply of nitrogen from animal manure exceeds 100 kg/ha in Belgium, Denmark, parts of Germany (e.g. Niedersachsen), Spain (Galicia and Cantabria), France (Bretagne), Luxembourg and the Netherlands.

Current nitrate policies in the EU are based on Council Directive 91/676/EEC on the protection of waters against pollution caused by nitrates from agricultural sources (Nitrates Directive). The Directive stipulates that Member States must implement the following provisions:

- All waters must be monitored, and zones vulnerable to nitrate leaching must be identified.
- Member States must establish Codes of Good Agricultural Practice.
- An Action Programme must be formulated in respect of the designated vulnerable zones.

As already underlined by Dosi and Zeitouni (Chapter 6, this volume), the directive is still awaiting full implementation, and a 1998 report from the European Commission has made it clear that most member states have failed to implement it adequately. Meanwhile, the Commission has begun legal procedures against several Member States because the degree of implementation has been inadequate and/or the directive incorrectly applied.

The directive prohibits compensation being paid to farmers for observing the required standards within the nitrate vulnerable zones. However, member states can and do offer a low proportion of grant aid to assist farmers in nutrient budget preparation and farm waste planning, and also in making capital investments to adjust their waste and fertiliser usage practices so as to facilitate compliance with the directive.

Essentially, the nitrate directive requires farmers to take remedial actions so as to reduce or avoid environmental harm until a specific environmental target is reached. According to the polluter pays principle, such costs are born by the perpetrator of such harm. Measures under Regulation 2078/92 on the introduction and maintenance of agricultural production methods compatible with the requirements that the environment be preserved and the countryside managed do however allow for the provision of compensation to farmers adopting practices which are more sensitive environmentally and provide positive environmental effects beyond environmental objectives.

13.2.2. *Pesticides*

Pesticides can have adverse impacts on human health and the environment. Monitoring programmes generally find levels over 0.1 μg/l in between 5 and 25% of samples in regions with intensive arable production and horticulture (including northern France and southern England). About 850 active substances are authorized for use in the EU, although only a small proportion of these are used on a large scale.

In 1996, the total use of pesticides in the EU was around 300 million kg of active ingredients. Pesticide sales fell by 13% during the first half of the 1990s. The reduction in total quantities of pesticides used is mainly due to the use of new compounds requiring lower dosages, farm management changes (for example, the application of integrated pest management strategies), national mandatory measures (for example, in Denmark, the Netherlands, Finland and Sweden) and Community legislation which establishes limits for pesticide residues. Figures from Eurostat on pesticide sales indicate that the largest reductions were achieved during the early 1990s in countries with targeted pesticide reduction programmes, such as Finland, the Netherlands, Denmark and Sweden (Europe Environment, 1999).

Pesticide use figures went up again by 6% overall in 1996, and markedly in some countries (e.g. Spain, France and the UK). In the case of Spain, this may have been largely due to the ending of a period of several years of drought. On a per hectare basis, the use of pesticides is low in Denmark, Spain, Ireland and Portugal, and highest in Belgium and the Netherlands.

Reform of the arable cropping system in 1992 has only contributed to a reduction of about 3% of total pesticide use since then (Falconer and Oskam, 2000). This reform has contributed to a reduction of around 10% in pesticide use in growing arable crops, including a reduction due to set-aside requirements and a reduction as a consequence of lower intervention prices for cereals, oilseed and protein crops.

The main legislative constraints are contained in Council Directive 91/414/EEC of 15 July 1991 on the placing of EEC-accepted plant protection products on the market. Its main goal is to lay down uniform rules in respect of the conditions and procedures for authorising plant protection products. In addition to this authorization procedure, this Directive also concerns the placing of pesticides on markets within the Community, and use and control of them. This Directive is applicable to the whole territory of the European Union. It is essentially intended to ensure that authorized products are effective and have an acceptable impact on the environment if they are used properly.

The basic principles of this Directive include the following:

- the development of a Community list of permissible active substances which are considered to be acceptable for human or animal health or the environment,
- a review programme for existing active substances, with uniform rules on the conditions and procedures for authorization of pesticides by Member States,
- authorization by Member States of individual plant protection products, which for new active substances or reviewed active substances may only contain those included on the positive list (with uniform principles to be the common criteria[1]),
- mutual recognition of acceptance by Member States, provided that the plant

health, agricultural and environmental conditions are comparable in the regions concerned,

- harmonized rules on classification, packaging and labelling.

The substitution principle is not included in Directive 91/414/EEC. It is, however, included in Council Directive 98/8/EC of February 16, 1998 on the placing of biocidal products on the market (Biocides Directive). The main objective of this directive is to harmonize the registration of non-agricultural pesticides and other biocides within the EU. Essentially, this principle requires taking steps to avoid using products for which less harmful substitutes are available.

Currently, the use of pesticides is mainly subject to control at national and sub-national levels. For example, mandatory inspection of spraying equipment is applied in several Member States. The use of pesticides is also prohibited or severely restricted in environmentally sensitive areas (e.g. water catchment areas) in most of the EU, as well as along streams and lakes. Strict rules on the use of pesticides are also applied in groundwater protection areas. In several countries, there are cost-sharing programmes for farmers facing strict rules limiting pesticide use along watercourses.

13.2.3. *Agricultural water abstractions*

It is primarily in areas where water supplies are increasingly being limited that irrigated agriculture is an important part of agriculture. Agricultural products deriving from irrigated land contribute a far greater proportion of total production compared to the total land area. About 15% of cropland in the USA is irrigated, but this contributes almost 40% of total production value from crops (ERS, 1994). Agriculture is the single most significant user of water in Greece (80% of all water abstractions), Italy and Portugal (around 50%) and Spain (65%). Average rates of use may exceed 7000 m^3/ha of irrigated land in Spain, and around 3000 m^3/ha of irrigated land in France. For Europe as a whole, about a quarter of the abstracted water is used in agriculture.

The highest proportion of irrigated land is found in the southern Member States and the south of France, and a significant proportion of horticultural and other cropland in the drained countries of the Netherlands, southern Denmark and Flanders (Belgium) are also irrigated during the dry summer season. Crops such as potatoes, salad vegetables, soft fruit and sugarbeet are particularly prone to damage from drying-out in summer, so irrigation, drawn particularly from groundwater sources, is frequently used. As a consequence, the water table is being lowered in many areas and this is leading to problems of salinization similar to those found in Mediterranean areas. Pollution, salinization and over-extraction in the aquifers are creating severe environmental pressures in Spain.

Irrigated land commonly contributes to a large share of agricultural production. In Spain, irrigated agriculture accounts for about 15% of cultivated land but it generates about 60% of total agricultural production and an even higher share (80%) of total agricultural produce exports (Varela-Ortega and Sumpsi, 1998).

In several Member States, abstraction licences are granted with associated requirements. For example, a licence may be given to abstract from a river during the wetter part of the year only, on condition that the water is then stored on the farm in a reservoir with a certain maximum capacity for use later in the year. In addition, the licence may specify what techniques of irrigation are to be used (for example, drip or trickle systems), although this is more commonly left for farmers to decide, using state-funded extension services which advise on how to improve the cost-efficiency of irrigation systems.

13.3. The role of CAP

Technological developments, high prices of land in large areas of Europe and the economic considerations of maximising returns and reducing costs have been the main driving forces towards intensification of agricultural land. CAP has also played a role in contributing to the intensification of agriculture. In attempts to better understand the environmental impacts of CAP, one of the most difficult tasks is probably isolating the impact of such a policy from other factors. Other factors of importance include policies at regional, national and Community levels (economic, environmental, fiscal and employment policies), trends in world market conditions, technological innovations and specific local factors.

13.3.1. *Market conditions*

Market conditions, balancing supply and demand, are vital to understanding the interactions between CAP and the environment. The 1992 reform of the arable cropping system, which gradually reduced intervention prices did not, for example, lead to a large decrease in market prices received by farmers in the UK. This was mainly because of the trend in world supplies and prices combined with the devaluation of the pound sterling. Nor was there an extensification of the production methods applied and there were few or no benefits to the environment from reduced inputs (Winter, 2000).

13.3.2. *Technological progress*

Market and price support measures under CAP will probably contribute to an increase in productivity, mainly through technological change and the transformation of agriculture from a labour-intensive industry to capital-intensive industry. Major progress has been achieved through the development and adoption of new production processes and altered pollution abatement processes such as drip irrigation control systems and low-emission manure application equipment.

Technological innovations in developing new pesticides have been vital to the reduction achieved during the past ten years. The reforms of the arable cropping system have probably only contributed to a limited reduction in pesticide use.

13.4. Identification of research gaps and research needs

Key driving forces for change in agricultural practices include world market conditions and consumer behaviour, technological development and world market conditions. The existing knowledge on the linkages between the CAP and the environment remains partial. Defining a research agenda is far from simple when there are many diverse and difficult issues to be addressed. Various factors need to be considered in shaping future European agricultural policy:

- the accession of central and eastern European countries to the EU. CAP needs to be adapted to this accession;
- the next round of multilateral trade negotiations. The EU and several other countries want to place consumer and health-related issues high on the agenda. The provision of support to farmers to encourage beneficial environmental and landscape effects will probably be high on the agenda as well.
- prospective trends in global food demand: with a possible doubling of world population and consumers demanding a broad and diverse package of agricultural produce, there may potentially be big changes;
- the development of new technologies (information technology, biotechnology);
- the changes in consumer preferences and public concerns, with consumers wanting a broader and more diverse range of produce with food quality, health and animal welfare considerations at the fore;
- a broadening of the environmental agenda for agricultural policy, including landscape, atmospheric pollution, soil conservation, biodiversity and the rational use of water resources;
- the diminishing size of the agricultural sector, and involvement of other actors who might also contribute to providing environmental goods and services.

A recent investigation of research on the environmental impacts of CAP identified some key research features (Brouwer and Lowe, 2000).

- First, there is a northern bias in the research coverage, with the majority of studies in Germany, the Netherlands and the UK, and much fewer for southern Europe. Research also puts stronger emphasis on temperate rather

than Mediterranean crops, and on intensive rather than extensive production systems, with water pollution problems caused by pesticides and nitrates reasonably well covered but not water supply problems and over-abstraction.
- Second, there is a strong emphasis put on research into agri-environmental measures compared to the other elements of CAP. These measures still account for only around 4% of the overall CAP budget, remain a minor component of CAP, and may draw attention away from the bigger picture.
- Third, there is limited if any effort being put into the environmental effects of certain commodity production systems (for example, tobacco and sugar), horizontal socio-structural measures (including less favoured areas), regional and rural policies and other CAP measures such as incentives for alternative crops, quality and label policy, and farm diversification.
- Fourth, there is a tendency for single country studies focusing on specific policy measures and single disciplinary approaches to be made. Integrated studies focussing on specific regions or farming systems remain limited.

Policy is tending to become more decentralized, with greater emphasis given to subsidiarity. It implies that there is an increasing need for policy evaluations. In order to fill these gaps, a series of research needs have been formulated (Brouwer and Lowe, 2000; Crabtree and Brouwer, 1999).

13.4.1. *A clear vision of the future role of European agriculture is required*

There is a need to improve the diversity of farming across Europe, including the role of different types of farming in their specific rural contexts. The policy framework needs to be sensitive to variability within agriculture and its multi-functionality. Commercial farming is likely to be subject to market forces and environmental regulation, while support should be given to non-commercial farming to supply services beneficial to the public domain, such as environmental quality, landscape maintenance, resource management and the provision of nature. The implications of such a dual structure and efforts to rebalance the forces of intensification and concentration call for careful analysis and better information.

Generally speaking, there is a need to better understand the implications of farming diversity across Europe, including the role of different types of farming in their specific rural contexts. This could largely contribute to providing different options for the policy community. There is thus a need for a systematic comparative regional geography of farming systems and their environmental rural development relationships. Environmental indicators, preferably in a regional context, are going to be very important in the move from a sector-based approach to a more territorial policy.

13.4.2. *There is a need to examine instruments for the integration of agriculture and environment*

Environmental concerns about agriculture can be observed at a local, regional, national, continental and global scale. In the context of CAP, there is considerable debate about its role in alleviating pressures on the environment and enhancing beneficial effects. CAP measures alone, though, would not achieve environmental targets, but must complement and be complemented by environmental measures and regulations. To understand what the best combinations might be, we need to know more about farmers' reactions and strategies when they are confronted with environmental policies/regulations. The efficiency and effectiveness of payments and of different instruments needs to be assessed on the basis of a clearer European framework specifying the principles for a division of labour between payments and regulation related to the positive and negative externalities of agricultural production. As part of this, the different environmental standards that farming faces in different countries need to be correlated, with a view to considering how to harmonize them. Countries, however, differ in their demand and supply of amenities, commodities and services because of differences in people's preferences and natural and environmental resource features.

Agri-environmental measures under Regulation 2078/92 are an important element of compensatory payments for the provision of services and goods by rural societies. Some measures are well targeted, while others are far more general. Co-financing of the measures may have to be enlarged by means of providing additional finance, wider eligibility, and regional level programming. This will call for more effective specification, monitoring and assessment of the benefits to be achieved.

In the gap between minimum environmental standards and the sort of environmental services for which payments are made, there is a range of societal expectations about the responsibilities that farmers should undertake on behalf of the environment. These are expressed in a range of Codes of Good Agricultural Practice and in the rising demand to add environmental conditions to the provision of compensatory payments to farmers. Research could contribute to the former development by examining the consistency of methods applied, and reviewing how codes operate in different countries and with what results.

In order to arrive at a concrete formulation of the environmental and conservation conditions that have to be fulfilled, a proper definition of the term Good Agricultural Practice (GAP) is essential. GAP may then become a benchmark used to decide whether a farmer is or is not eligible for income support (for example, in the context of 'cross compliance') (Baldock and Mitchell, 1995). Member States could currently take environmental measures that suit the specific situation in the country concerned. In fact, the Netherlands is currently considering putting environmental constraints on the provision of compensatory payments for maize. It is expected that this system will be introduced in 2000. The agri-environmental measures offer financial support to farmers who provide environmental services on a voluntary and contractual basis or ensure improvement in the environmental performance of their production techniques.

13.4.3. *Greater attention paid to link economic activities to environmental quality*

Several analytical frameworks have been developed to address the impact of agricultural policies on the environment. A very widely used framework has been developed within what is called a driving force (state) response framework (DSR). The driving forces mainly include farming practices which induce changes in the state of the environment. The state refers to the environmental conditions arising from these driving forces, while the responses refer to the reactions by the actors involved (for example, farmers who may adopt new practices). The DSR framework has been taken up by Eurostat, OECD and the EEA in their efforts to integrate driving forces into the state of the environment. The DSR Framework has been broadened into the concept of DPSIR, which includes causal links made between driving forces and pressures on environmental states (P) and the impacts (I) on human activities, and subsequently leading to political responses (R). Integrated environmental assessments (IEA) such as this have been identified as priority areas of the European Environment Agency, but current experiences remain scarce. One of the few examples of application of IEA has been on eutrophication of water, examined for Denmark (Iversen *et al.*, 1997).

13.4.4. *Greater attention paid to market conditions*

European farmers mainly respond to the market conditions for their commodities, both in terms of availability and prices, but also in terms of exchange rates between national currencies. For example, the market conditions for raw materials to produce compound feed play a very significant role in global trade patterns. The price of soybean products may show large inter-annual variations. There are many factors which might contribute to an increase in soybean prices, including poor harvests in other parts of the world, rising US dollar exchange rates, and global population trends. For example, the high prices of soybean products during the end of 1997 was considered to be partly due to lower fish catches in Latin America, which increased the demand for products like soy, because of the need for protein in food. The import of soybean (in the form of soybean products) is very responsive to such price variations, and this also affects the composition of compound feed (mainly its protein content) and the subsequent consequences of this on the nitrogen excreted by livestock.

13.4.5. *Greater attention paid to market initiatives to support sustainable agricultural production methods*

Consumers in the EU are increasingly concerned about the quality of agricultural products (for example, human health concerns related to residues in food) and the production methods applied: are they sustainable production methods? This trend towards public concern is leading to consumer demand for more information on products, their provenance and the production methods applied. Food labelling can supply the desired information (or at least part of it), but other information transfer mechanisms exist. To comply with these public and consumer concerns, private firms in the agrifood system are venturing into the marketing of products which have high quality standards. In recent years, the number of these private market ventures has increased, particularly by retailers, but also by farmers, by wholesalers and by food processors (Van der Grijp and den Hond, 1999). To obtain the desired quality attributes, farmers need to comply with product and process standards set by their clients (the food processors, wholesalers and retailers). As a result, farmers often need to change their production methods, including the purchase of other inputs. Organic production methods have increased as a result of the rising demand for organic food, the active involvement of retailers and higher prices at the farm gate. Other market efforts have also contributed to more environmentally friendly production methods. In the Netherlands, for example, Milieu Project Sierteelt (MPS) started up in 1993 as a system to improve the environmental profile of flowers and ornamental plants. It was initiated by the flower auctions in co-operation with primary producers' organizations and currently has a share of half of the land used for growing flowers and ornamental plants.

The objectives of MPS are to provide incentives to reduce pressures on the environment, and to develop instruments for improving the environmental profile of the sector. The factors involved are use of energy, pesticides and nutrients, both in absolute terms, as well as in terms of means of application. Its accounting system also keeps records about the amount of waste supplied by the producers. Environmental profiles are developed on a periodic basis (the performance of individual holdings during a period of 13 weeks) (Brouwer and van Bruchem, 1999).

When it started in 1994, about 1000 holdings joined the system. This has gradually increased to a current participation of 3300 registered holdings, and almost 2900 holdings classified within the accounting system. On holdings that have been in the system since 1995, the use of pesticides has been reduced by some 25% (in kg of active ingredients per hectare). A similar reduction has been achieved in the use of phosphate fertilisers, while the use of nitrogen fertilisers has been reduced by some 12%. Inter-annual changes of energy consumption are large, which partly reflects temperature variation.

13.4.6. *Greater attention paid to the evaluation of policy measures and data needs*

Procedures for monitoring and evaluation of the environmental impacts of policy measures are essential for assessing the efficiency and effectiveness of such measures.

High priority should be given to improving the information available to policy makers, with a focus on describing the effects of agriculture on water quantity and water quality, and including a range of issues, such as driving forces, state of the environment and policy responses. Policy analysis needs to be backed up with up-to-date and high quality data. Data sets that are consistently defined and collected and allow for cross-national comparison are urgently needed.

Notes

1. The Uniform Principles are established by Council Directive 97/57/EC as Annex VI of Directive 91/414/EEC, and include criteria regarding ecotoxicity, human toxicity, environmental fate, packaging and labeling.

References

Baldock, D., Mitchell, K., 1995, Cross-Compliance within the Common Agricultural Policy: A Review of Options for Landscape and Nature Conservation, London, Institute for European Environmental Policy (IEEP), 83 pp.

Brouwer, F., van Bruchem, C., 1999, Landbouw, Milieu en Economie. Editie 1998, The Hague, Agricultural Economics Research Institute, Rapport 6.98.97.

Brouwer, F., Lowe, P., 2000, CAP and the environment: policy development and the state of research, in Brouwer, F., Lowe, P., eds., CAP Regimes and the European Countryside: Prospects for Integration Between Agricultural, Regional and Environmental Policies, Wallingford, CAB International, pp. 1–14.

Crabtree, B., Brouwer, F., 1999, Discussion and conclusions, in Brouwer, F., Crabtree, B., eds., Environmental Indicators and Agricultural Policy, Wallingford, CAB International, pp. 279–285.

ECPA 1996, European Crop Protection: Trends in Volumes Sold, 1985–1995, Brussels, European Crop Protection Association.

EEA, 1999, Environment in the European Union at the Turn of the Century, European Environment Agency, Copenhagen.

ERS, 1994, Agricultural Resources and Environmental Indicators, Washington, United States Department of Agriculture, Agricultural Handbook Number 705.

European Commission, 1995, Progress Report on the European Community Programme of Policy and Action in Relation to the Environment and Sustainable Development, Communication of the European Communities, Brussels, COM(95)624.

European Commission and Eurostat, 1999, Agriculture, Environment, Rural Development: Facts and Figures – A Challenge for Agriculture, DG VI, DG XI and Eurostat, Brussels.

European Economy, 1997, Towards a Common Agricultural and Rural Policy for Europe, Directorate-General for Economic and Financial Affairs, Reports and Studies, Vol. 5, Brussels.

Europe Environment, 1999, Pesticides: EU sales on the rise again, Brussels, Europe Environment, Vol. 537, January 19, 1999.

Falconer, K., Oskam, A., 2000, The arable crops regime and the use of pesticides, in Brouwer, F., Lowe, P., eds., CAP Regimes and the European Countryside: Prospects for Integration Between Agricultural, Regional and Environmental Policies, Wallingford, CAB International, pp. 87–102.

van der Grijp, N.M., den Hond, F., 1999, Green Supply Chain Initiatives in the European Food and Retailing Industry, Amsterdam, Free University, Institute for Environmental Studies.

Hoogervorst, N.J.P., 1996, Moving models and pragmatic policies to reduce agricultural nutrient emissions in the Netherlands, in Walter-Jørgensen, A., Pilegaard, S., eds., Integrated Environmental and Economic Analysis in Agriculture, Copenhagen, Statens Jördbrugs- og Fiskeriøkonomiske Institut (SJFI), pp. 247–274.

Iversen, T.M., Kjeldsen, K., Kristensen, P., de Haan, B., van Oirschot, M., Parr, W., Lack, T., 1997, Integrated Environmental Assessment on Eutrophication: A Pilot Study, Silkeborg, National Environmental Research Institute, NERI Technical Report, No. 207.

Stanners, D. Bourdeau P., eds., 1995, Europe's Environment – The Dobrís Assessment, A report of the European Environment Agency, Copenhagen, Denmark.

Varela-Ortega, C., Sumpsi, J., 1998, Spain, in Brouwer, F., Lowe, P., eds., CAP and the Rural Environment in Transition: A Panorama of National Perspectives, Wageningen, Wageningen Pers, pp. 201–240.

Winter, M., 2000, The Arable Crops Regime and the Countryside Implications, in Brouwer, F., Lowe, P., eds., CAP Regimes and the European Countryside: Prospects for Integration between Agricultural, Regional and Environmental Policies, Wallingford, CAB International, pp. 117–136.

Index

FONDAZIONE ENI ENRICO MATTEI (FEEM) SERIES ON ECONOMICS, ENERGY AND ENVIRONMENT

This series serves as an outlet for the main results of FEEM's research programmes in the areas of economics, energy and environment.

---

1. C. Carraro and D. Siniscalco (eds.), *The European Carbon Tax: An Economic Assessment.* 1993 ISBN 0-7923-2520-6
2. C. Carraro (ed.), *Trade, Innovation, Environment.* 1994 ISBN 0-7923-3033-1
3. C. Dosi and T. Tomasi (eds.), *Nonpoint Source Pollution Regulation: Issues and Analysis.* 1994 ISBN 0-7923-3121-4
4. C. Carraro, Y. Katsoulacos and A. Xepapadeas (eds.), *Environmental Policy and Market Structure.* 1996 ISBN 0-7923-3656-9
5. C. Carraro and A. Haurie (eds.), *Operations Research and Environmental Management.* 1996 ISBN 0-7923-3767-7
6. I. Musu and D. Siniscalco (eds.), *National Accounts and the Environment.* 1996 ISBN 0-7923-3741-7
7. C. Carraro and D. Siniscalco (eds.), *Environmental Fiscal Reform and Unemployment.* 1996 ISBN 0-7923-3750-6
8. A. Beltratti, *Models of Economic Growth with Environmental Assets.* 1996 ISBN 0-7923-4032-9
9. G. Chichilnisky, G. Heal and A. Vercelli (eds.), *Sustainability: Dynamics and Uncertainty.* 1998 ISBN 0-7923-4698-X
10. R. Roson and K.A. Small (eds.), *Environment and Transport in Economic Modelling.* 1998 ISBN 0-7923-4913-X
11. A. Markandya and M. Pavan (eds.), *Green Accounting in Europe - Four Case Studies.* 1998 ISBN 0-7923-5470-2
12. A. Lanza (ed.), *Resources Accounting in China.* 1999 ISBN 0-7923-5485-0
13. C. Carraro (ed.), *International Environmental Agreements on Climate Change.* 1999 ISBN 0-7923-5515-6
14. C. Carraro and F. Lévêque (eds.), *Voluntary Approaches in Environmental Policy.* 1999 ISBN 0-7923-5516-4
15. C. Carraro (ed.), *Efficiency and Equity of Climate Change Policy.* 2000 ISBN 0-7923-6262-4
16. I. Musu (ed.): *Sustainable Venice: Suggestions for the Future.* 2000 ISBN 0-7923-6458-9
17. C. Dosi (ed.): *Towards Integration between Agricultural Policy and Water Resources Management.* 2001 ISBN 0-7923-6805-3

KLUWER ACADEMIC PUBLISHERS – DORDRECHT / BOSTON / LONDON